THERMODYNAMICS AND INTRODUCTORY STATISTICAL MECHANICS

THERMODYNAMICS AND INTRODUCTORY STATISTICAL MECHANICS

BRUNO LINDER
Department of Chemistry and Biochemistry
The Florida State University

WILEY-INTERSCIENCE
A JOHN WILEY & SONS, INC. PUBLICATION

Published by John Wiley & Sons, Inc., Hoboken, New Jersey.
Published simultaneously in Canada.

For general information on our other products and services please contact our Customer Care Department
within the U.S. at 877-762-2974, outside the U.S. at 317-572-3993 or fax 317-572-4002.

Wiley also publishes its books in a variety of electronic formats. Some content that appears in print,
however, may not be available in electronic format.

Library of Congress Cataloging-in-Publication Data:

Linder, Bruno.
   Thermodynamics and introductory statistical mechanics/Bruno Linder.
      p.   cm.
   Includes bibliographical references and index.
   ISBN 0-471-47459-2
   1. Thermodynamics.   2. Statistical mechanics. I Title.
   QD504.L56 2005
   541'.369–dc22   2004003022

Printed in the United States of America

10 9 8 7 6 5 4 3 2 1

To Cecelia
...and to William, Diane, Richard, Nancy, and Carolyn

CONTENTS

APPENDIX II   SOLUTIONS TO PROBLEMS   177

INDEX   201

PREFACE

This book is based on a set of lecture notes for a one-semester first-year chemistry graduate course in *Thermodynamics and Introductory Statistical Mechanics*, which I taught at Florida State University in the Fall of 2001 and 2002 and at various times in prior years. Years ago, when the University was on the quarter system, one quarter was devoted to Thermodynamics, one quarter to Introductory Statistical Mechanics, and one quarter to Advanced Statistical Mechanics. In the present semester system, roughly two-thirds of the first-semester course is devoted to Thermodynamics and one-third to Introductory Statistical Mechanics. Advanced Statistical Mechanics is taught in the second semester.

Thermodynamics is concerned with the macroscopic behavior of matter, or rather with processes on a macroscopic level. Statistical Mechanics relates and interprets the properties of a macroscopic system in terms of its microscopic units. In this book, Thermodynamics was developed strictly from a macroscopic point of view without recourse to Statistical-Mechanical interpretation, except for some passing references to fluctuations in the discussion of critical phenomena. Both Thermodynamics and Statistical Mechanics entail abstract ideas, and, in my opinion, it is best not to introduce them simultaneously. Accordingly, the first 12 chapters (Part I) deal exclusively with Thermodynamics; Statistical Mechanics is only then introduced.

Thermodynamics, unlike some other advanced subjects in Physical Chemistry, does not require complicated mathematics, and for this reason the subject is often thought to be "easy." But if it is easy, it is *deceptively easy.* There are subtleties and conceptual difficulties, often ignored in elementary treatments, which tend to obscure the logical consistency of the subject. In this book, conceptual difficulties are not "swept under the rug" but brought to the fore and discussed critically. Both traditional and axiomatic approaches are developed, and reasons are given for presenting alternative points of view. The emphasis is on the logical structure and generality of the approach, but several chapters on applications are included. The aim of the book is to achieve a balance between fundamentals and applications.

In the last four chapters of the book, which are devoted to Statistical Mechanics, not much more can be hoped to be accomplished than to cover, from an elementary point of view, the basics. Nonetheless, for some students, especially those who are not physical-chemistry majors, it is essential that simple statistical-mechanical applications be included, thus acquainting students with some working knowledge of the practical aspects of the subject. Among the applied statistical-mechanical topics are numerical calculations of entropy and other thermodynamic functions, determination of equilibrium constants of gases, and determination of heat capacity of solids.

Although all fundamental equations are developed from first principle, my treatment is more advanced than what some students are likely to have been exposed to in elementary discussions of thermodynamics. This book is designed as a one-semester course, useful both to students who plan to take more advanced courses in statistical mechanics and students who study this as a terminal course.

An essential feature of this book is the periodic assignment of homework problems, reflecting more or less the content of the topics covered. Ten typical problem sets are included in Appendix I and their solutions in Appendix II.

I am grateful to Kea Herron for her help in formatting the manuscript, and to members of the Wiley Editorial staff, especially Amy Romano and Christine Punzo, for their advice, patience, and encouragement.

BRUNO LINDER

CHAPTER 1

INTRODUCTORY REMARKS

Thermodynamics, as developed in this course, deals with the macroscopic properties of matter or, more precisely, with processes on a macroscopic level. Mechanics (especially quantum mechanics) is concerned with molecular behavior. In principle, and in some limited cases, the molecular properties can be calculated directly from quantum mechanics. In the majority of cases, however, such properties are obtained from experimental studies such as spectral behavior or other devices, but the interpretation is based on quantum mechanics. Statistical mechanics is the branch of science that interconnects these seemingly unrelated disciplines: statistical mechanics interprets and, as far as possible, predicts the macroscopic properties in terms of the microscopic constituents.

For the purposes of the course presented in this book, thermodynamics and statistical mechanics are developed as separate disciplines. Only after the introduction of the fundamentals of statistical mechanics will the connection be made between statistical mechanics and thermodynamics. As noted, the laws of (macroscopic) thermodynamics deal with *processes* not *structures*. Therefore, no theory of matter is contained in these laws. Traditional thermodynamics is based on common everyday experiences. For example, if two objects are brought in contact with each other, and one feels hotter than the other, the hotter object will cool while the colder one will

Thermodynamics and Introductory Statistical Mechanics, by Bruno Linder
ISBN 0-471-47459-2 © 2004 John Wiley & Sons, Inc.

heat up. Because thermodynamics is based on the common experience of macroscopic observations it has a generality unequaled in science. "Classical Thermodynamics," Einstein remarked, "... is the only physical theory of universal content ... which ... will never be overthrown" (Schilpp, 1949).

1.1   SCOPE AND OBJECTIVES

Class make-up varies greatly. Some students take this course as part of one-year course, in preparation for a comprehensive or preliminary exam, required for a Master's or Ph.D. degree. Others sign up because they heard it was a "snap" course. Still others take it because they think, or their major professor thinks, that it may help them in their research. A course designed to satisfy all students' aspirations is difficult, if not impossible. A suitable compromise is one, which provides a reasonable balance between *fundamentals* and *applications*, which is the aim of this book.

1.2   LEVEL OF COURSE

Most students are likely to have had previous exposure to thermodynamics in some undergraduate course, such as physical chemistry, physics, or engineering. The present course is intended to be more advanced from the standpoints of both *principles* and *applications*. The emphasis is on the logical structure and generality of the subject. All topics of interest cannot possibly be covered in a semester course; therefore, topics that are likely to have been adequately treated in undergraduate courses are skipped.

1.3   COURSE OUTLINE

The idea is to proceed from the *general* to the *particular.* The following outline suggests itself.

Part I: Thermodynamics

>   A. *Fundamentals*
>   1. Basic concepts and definitions
>   2. The laws of thermodynamics
>   *2.1 Traditional approach*
>   *2.2 Axiomatic approach*
>   3. General conditions for equilibrium and stability

B. *Applications*
   1. Thermodynamics of (Real) gases, condensed systems
   2. Chemical equilibrium
      2.1 Homogeneous and heterogeneous systems
      2.2 Chemical reactions
   3. Phase transitions and critical phenomena
   4. Thermodynamics of one- and two-dimensional systems
      4.1 Film enlarging
      4.2 Rubber stretching

Part II: Introductory Statistical Mechanics

A. *Fundamentals*
   1. Preliminary discussion
   2. Maxwell-Boltzmann, Corrected Maxwell-Boltzmann Statistics
   3. Partition Functions
   4. Thermodynamic connection
B. *Applications*
   1. Ideal gases
   2. Ideal solids
   3. Equilibrium constant
   4. The bases of chemical thermodynamics

In addition, mathematical techniques are introduced at appropriate times, highlighting such use as:

1) Exact and inexact differentials (Section 3.3)
2) Partial Derivatives (Section 3.6)
3) Pfaffian Differential Forms (Section 4.6)
4) Legendre Transformation (Section 5.1)
5) Euler's Theorem (Section 5.7)
6) Combinatory Analysis (Section 13.5)

1.4 BOOKS

Because of the universality of the subject, books on Thermodynamics run into the thousands. Not all are textbooks, and not all are aimed at a particular discipline, such as chemistry, physics, or engineering. Most elementary chemical texts rely heavily on applications but treat the fundamentals lightly. Real systems (real gases, condensed systems, etc) are often not treated in any detail. Some books are strong on fundamentals but ignore applications.

Other books are authoritative but highly opinionated, pressing for a particular point of view.

Two chemical thermodynamics books, which discuss the fundamentals in depth, are listed below.

1. J. de Heer, *Phenomenological Thermodynamics*, Prentice-Hall, 1986.
2. J. G. Kirkwood and I. Oppenheim, *Chemical Thermodynamics*, McGraw-Hill, 1961.

Other books that may provide additional insight into various topics are listed in the Annotated Bibliography on page. . . .

PART I

THERMODYNAMICS

CHAPTER 2

BASIC CONCEPTS AND DEFINITIONS

Chapter 2 lists some of the basic concepts used in this book. Other, more difficult, concepts will be introduced as needed.

There are several approaches to the formulation of (macroscopic) thermodynamics. Most common is the *phenomenological* approach, based on observation. A phenomenological theory is one in which initial observations lead to a *law*. The law, in turn predicts phenomena, which can be verified by experimentation. Laws are seldom, if ever, formulated in terms of primary measurements because such formulations could lead to cumbersome statements. Rather, concepts are introduced that, as a result of the primary measurements, behave in a characteristic way, giving rise to new concepts, in term of which laws are expressed concisely and take on a simple form. I referred earlier to the common situation in which a hotter object in contact with a colder could only cool and not heat up; in a more precise language, this leads to the concept of entropy, as we shall see. The Second Law of Thermodynamics is then concisely formulated in terms of the entropy concept.

Laws can be used to *predict* events not yet measured. However, predictions should not be extrapolated *far beyond the domain* in which the primary measurements have been made. For example, extending the laws of thermodynamics obtained from measurements in a *macro system* to a *micro system* may lead to erroneous conclusions.

Thermodynamics and Introductory Statistical Mechanics, by Bruno Linder
ISBN 0-471-47459-2 © 2004 John Wiley & Sons, Inc.

2.1   SYSTEMS AND SURROUNDINGS

A *system* is part of the physical world in which one is interested. What is not the system is the environment or the *surroundings*. We distinguish between several types of systems:

1) An *isolated system* is a system that is totally uninfluenced by the surroundings. There is no possibility of exchange of energy or matter with surroundings.
2) A *closed system* is a system in which energy but not matter can exchange with the surroundings.
3) An *open system* is a system in which both energy and matter can exchange with the surroundings.

Isolated or closed systems are often referred to as *bodies.* Theorems will first be developed for isolated and closed systems and later generalized to open systems.

2.2   STATE VARIABLES AND THERMODYNAMIC PROPERTIES

For a complete description of a macroscopic body, it is not enough to specify the identity of the substance. The *state* of the system must also be specified. The state is completely defined by the values of the thermodynamic properties or thermodynamic variables of the system. Thermodynamic properties (such as temperature, pressure, amount of substance, energy, etc.) are properties that do not depend on the rate at which something happens. For example, electric current and thermal conduction are rates and not thermodynamic variables. State variables are fully determined by the values at *present* and do not depend on the *previous* history of the system. In general, not all variables need to be specified to define the state of the system because the variables are interdependent and only a small number can be varied independently. These are referred to as *independent* variables. The rest are *dependent* variables.

The number of independent variables needed to describe a macroscopic system thermodynamically is generally small, perhaps consisting of half a dozen or so variables. On the other hand, there is an underlying atomic structure (see Chapters 13–16 on statistical mechanics), which requires something on the order of a multiple of Avogadro's number of coordinates to fully describe the system mechanically. How is it then that we can characterize the macroscopic behavior thermodynamically in terms of a very small number of variables? The answer is that in a macroscopic observation only averages of atomic coordinates are observed. This has the effect of eliminating *most*, but *not all*, coordinates. For example, suppose we want to measure the length of a side of a crystalline solid by placing it near a ruler and taking snapshots with a high-speed camera. Even during these rapid measurements, atoms of the solid undergo billions of vibrations, which are averaged out in a macroscopic description. *The averaged length only survives.*

2.3   INTENSIVE AND EXTENSIVE VARIABLES

Intensive variables or properties are properties that are independent of the amount or mass of the material. *Extensive* properties depend on the mass. But there is more to it. If the system is divided into several parts, the value of the total extensive property must equal the sum of the values of the parts. Later (Chapter 5, Section 5.7), intensive and extensive properties will be discussed from a mathematical point of view.

2.4   HOMOGENEOUS AND HETEROGENEOUS SYSTEMS, PHASES

If the intensive variables are uniform throughout the system or if the variables change continuously (as air in a gravitational field), the system is said to be *homogeneous*. If some of the intensive properties change discontinuously, the system is said to be *heterogeneous*. A *phase* is a homogeneous subsystem. It is not necessary that all parts of a phase be contiguous. Ice chunks floating in water, for example, represent a two-phase system. A system that consists of several phases is obviously heterogeneous.

2.5   WORK

This concept is taken from electromechanics (mechanics, electricity, and magnetism) and is likely to be familiar from other studies. There are various kinds of work. All elements of work, dw, can be written as the product of a generalized force, X, and a generalized displacement, dx. That is, $dw = Xdx$ (or sum of terms like $\Sigma_i X_i \, dx_i$).

There is no general agreement regarding the sign of w. Some authors use the convention that w is positive if done by the system on the surroundings. Others prefer to take work to be positive if done by the surroundings on the system. *In this course, w is regarded as positive if done on the system by the surrounding.*

Examples of forms of work include the following.

- Pressure-volume or P-V work: $dw = -P_{ex}\,dV$, where P_{ex} is the external pressure and V the volume.
- Gravitational work: $dw = mgdh$, where m is the mass, g the gravitational constant, and h the height.
- Electrical work: $dw = \mathcal{E}dQ$, where $\mathcal{E}$ is the electric potential difference and Q the charge.
- Wire or rubber stretching: $dw = fL$, where f is the tensile force and L the length.
- Surface Enlargement: $dw = \sigma d\mathcal{A}$, where σ is the surface tension and $\mathcal{A}$ the area.

In all these examples, the generalized force is the external one (i.e., the force acting on the system). Only in reversible processes (to be discussed shortly) are the external and internal forces equal to each other.

2.6   REVERSIBLE AND QUASI-STATIC PROCESSES

Consider a gas in a cylinder fitted with a piston. If the gas is in equilibrium, its state is determined by a small number of macroscopic variables, such as pressure, volume, and composition. However, if the piston is in motion, the pressure varies from point to point. A pressure *tensor*, rather than a *scalar*, is needed to describe the motion. A rigorous treatment of pressure-volume work is obviously a problem of great complexity. However, there is one type of process that can be described simply, namely, a process that changes extremely slowly so that, for all practical purposes, the internal pressure is infinitesimally less than the external pressure (obviously an idealization!). The system will be effectively in equilibrium, and the internal pressure will be basically the same as the external pressure. The work can then be calculated by using the internal pressure, which is generally known (for example, from an equation of state). This type of transformation, introduced by Carathéodory (1909), is referred to as *quasi-static*. It cannot be realized exactly, but it may approximately represent the real situation in practice.

A concept closely related to the concept of quasi-static transformation is the concept of *reversibility*. Before discussing this concept in detail, let us inquire what some authorities have to say about the relation between the concept of reversible and quasi-static changes. de Heer (de Heer, 1986) quotes the following statements from well-established authors:

- "Quasi-static transitions are, in fact, reversible, but it is by no means obvious that all reversible transitions are quasi-static." [from H. A. Buchdahl]
- "Every reversible process coincides with a quasi-static one." [from H. B. Callen]
- "All reversible processes are quasi-static but the converse is not true." [from J. Kestin]

What appears to be the problem, as shown below, is that the terms *quasi-static* and *reversible* are not uniquely defined.

2.6.1 Quasi-Static Process

The most widely used definition of quasi-static process is the one due to Carathéodory (Carathéodory, 1909), which states that a "quasi-static process is one that proceeds infinitely slowly via a continuous succession of equilibrium states." The restriction to a continuous sequence of internal equilibrium state ensures that, for example, in a compression of a gas, the internal pressure is infinitesimally smaller than the external pressure. Obviously, this can be true only if there is no friction. Furthermore, by an infinitesimal change of the forces, the process can be reversed *along exactly the same path*. This means that, on completion of the reverse process, *the system is restored to its original values but so is the environment*. Restoration of the system to its original state as well as of the environment is an essential requirement (see Section 2.6.2 below) of a *reversible* process.

Another definition of quasi-static process stipulates that the change proceeds at an infinitesimally slow rate but not necessarily via a continuous succession of equilibrium states. Such a process cannot be reversed by an infinitesimal change of the forces. As an example, consider a system and its surroundings—initially at finite (not infinitesimally) different temperatures—to be connected by a poorly heat-conducting metal plate. The heat transfer will proceed infinitely slowly, and the process may be dubbed quasi-static—but it is not reversible. Once equilibrium has been established, it is impossible to restore both system and surroundings to their initial states without producing finite changes. Only if the initial temperature difference between system and surroundings is infinitesimally small, rather than finite, will the transformation be reversible or quasi-static in the Carathéodory sense.

2.6.2 Reversible Process

So, how does one define a reversible process?

One definition states that a reversible process *proceeds along a continuous sequence of internal equilibrium states so it can return along exactly the same path.* This implies that both system and surrounding can be restored to the initial values. From this point of view, reversible and quasi-static processes (á la Carathéodory) are the same.

Another definition states that a reversible process is one in which the system is taken from state A to state B and returned to state A, not necessarily along the same path, but along a path *such that there are changes in neither the system nor the surroundings.* This definition makes no reference to intermediate equilibrium states.

It has been suggested that the first definition of reversibility be called *retraceable* and the second *recoverable.* de Heer adds, "No one has proven that any recoverable process must be retraceable, nor has any one come up with an example where a recoverable process is not retraceable."

It is apparent now why seemingly contradictory statements are made regarding the relationship between reversible and quasi-static changes. The same words are used to describe different phenomena.

For the purposes of the course outlined by this book, quasi-static process will be defined as one that proceeds via a sequence of equilibrium states, and reversible change will be defined as a change that proceeds along a continuous sequence of equilibrium states; in addition, the concepts of quasi-static and reversible processes will be used interchangeably. The concept of reversibility is important, not only because it enables us, when appropriate, to carry out calculations that would otherwise be difficult, if not impossible, but also because reversibility (or the lack thereof) plays an essential role in establishing criteria for natural occurring or spontaneous processes, as will be shown later. *Spontaneous processes are irreversible.*

Note: When a transition is carried out under quasi-static conditions, the work done by the surroundings on the system is the maximum work (by our convention) because the internal pressure differs only infinitesimally from the external one. In an expansion, the work (again by our convention) is negative and as low as possible; that is, it is a minimum. Had we adopted the convention that work is positive when done on the surroundings, the reversible work would have been maximum in an expansion and minimum in compression. The absolute values of work are always maximum in a reversible change regardless of convention and regardless of whether the transition is an expansion or compression.

In an irreversible change, the external pressure in a compression has to be greater than the internal pressure by a *finite* amount, since not all work energy, $w = - \int P_{ext}dV$, is utilized to compress the gas but some is needed to overcome friction.

2.7 ADIABATIC AND DIATHERMAL WALLS

These concepts take on an important role in the axiomatic approach to thermodynamics.

Adiabaic wall or *adiabaic boundary* is one in which the state of the system can be changed *only* by moving the boundaries (i.e., doing mechanical work) or by applying an external field.

Diathermal wall or *diathermal boundary* is one in which the state of the system can be changed by means *other* than moving the boundary or by applying a field.

2.8 THERMAL CONTACT AND THERMAL EQUILIBRIUM

Two other concepts that will be needed to discuss the laws of thermodynamics, in particular the Zeroth and the First Law, are the concepts of thermal contact and thermal equilibrium.

Thermal contact refers to systems in contact via a diathermal wall. When objects are brought into thermal contact, the macroscopic properties may initially change but after some time no further changes will occur.

Thermal equilibrium refers to systems in thermal contact that do not change with time.

CHAPTER 3

THE LAWS OF THERMODYNAMICS I

All of us have some intuitive feeling of what is meant by temperature or heat, but defining them rigorously leaves much to be desired. The difficulties encountered in defining these concepts have undoubtedly contributed to the abandonment of the traditional approaches to thermodynamics in favor of axiomatic ones.

The traditional way to discuss temperature or heat is to first define one of these concepts and deduce the other from it. Ideally, the definitions should be in operational terms, that is, in terms of experimental procedures that can be measured. Attempts to define temperature in terms of heat are bound to cause difficulties. Normally, heat is not observed directly but inferred from changes in temperature. Statements such as radiant energy, thermal flow, heat flow, and so forth are sometimes used to define heat. The definition of temperature is contingent on knowing what *heat* is, which is here vaguely defined and not operationally. The more common, *traditional* approach is to define temperature first and heat afterward. In the *axiomatic* approach, the definition of heat is not contingent on knowing what temperature is.

It was once suggested that temperature be considered a primary quantity, like length and time, which cannot be analyzed into something simpler. This idea is unsatisfactory from the standpoint of statistical mechanics, which connects the *thermodynamic* properties to the *mechanical*, as we shall see

Thermodynamics and Introductory Statistical Mechanics, by Bruno Linder
ISBN 0-471-47459-2 © 2004 John Wiley & Sons, Inc.

later. If temperature were a primary quantity, which cannot be further simplified, the mechanical properties would also have to be described in terms of the temperature. They are not.

A critical discussion of temperature did not come about until the latter half of the 19th century. Maxwell made the definition of temperature contingent on an observation, now often referred to as the *Zeroth Law of Thermodynamics*.

3.1   THE ZEROTH LAW—TEMPERATURE

Zeroth Law states that if two bodies are in thermal equilibrium with a third, they are in thermal equilibrium with each other.

This statement leads directly to an operational definition of temperature, given below.

Let A be a test body (e.g., a thermometer). Then all bodies in thermal equilibrium with it are in thermal equilibrium with each other; that is, they have a property in common. This property is called temperature. We may characterize the states of all systems in thermal equilibrium by assigning a number. That number represents the temperature. (If this "proof" appears to be less than convincing, a mathematical analysis, given below, analyzes the concept in greater detail and shows that *temperature* can be expressed entirely in terms of the mechanical variables, such as pressure, volume, etc.)

Consider three systems: A, B, and C. Let the mechanical variables be P_A, P_B, P_C, $\overline{V}_A$, $\overline{V}_B$, and $\overline{V}_C$, where P is pressure and $\overline{V}$ is molar volume. If A and C are connected via an adiabatic wall, four variables $(P_A, \overline{V}_A, P_C, \overline{V}_C)$ will be needed to describe the composite system. On the other hand, if A and C are connected by a diathermal wall and are in thermal equilibrium, not all four variables will be independent. We can express this interdependency by writing

$$f_1(P_A, \overline{V}_A; P_C, \overline{V}_C) = 0 \quad \text{or} \quad P_C = \phi_1(\overline{V}_A, \overline{V}_C; P_A) \qquad (3\text{-}1)$$

Similarly, if B and C are in thermal equilibrium, we can write

$$f_2(P_B, \overline{V}_B; P_C, \overline{V}_C) = 0 \quad \text{or} \quad P_C = \phi_2(\overline{V}_B, \overline{V}_C; P_B) \qquad (3\text{-}2)$$

If A and B are in thermal equilibrium with C, then by Zeroth Law they are in thermal equilibrium with each other. We conclude that

$$f_3(P_A, \overline{V}_A; P_B, \overline{V}_B) = 0 \qquad (3\text{-}3)$$

implying that $P_A, \overline{V}_A, P_B$, and $\overline{V}_B$ are interdependent.

From *Eqs. 3-1* and *3-2*, we obtain

$$\phi_1(\overline{V}_A, \overline{V}_C; P_A) = \phi_2(\overline{V}_B, \overline{V}_C; P_B)$$

implying that a functional relation exists between P_A, $\overline{V}_A$, P_B, $\overline{V}_B$, *and* $\overline{V}_C$ or

$$f_3^*(P_A, \overline{V}_A; P_B, \overline{V}_B; \overline{V}_C) = 0 \qquad (3\text{-}4)$$

How can *Equations 3-3* and *3-4* be reconciled? *Equation 3-3* indicates that P_A, P_B, $\overline{V}_A$, and $\overline{V}_B$ are interdependent but do not depend on $\overline{V}_C$. *Equation 3-4* says that the same variables are interdependent but also that they are dependent on $\overline{V}_C$. The two equations can be reconciled if we assume that the functions ϕ_1 and ϕ_2 have the following general form:

$$\phi_1(\overline{V}_A, \overline{V}_C; P_A) = \theta_1(\overline{V}_A, P_A)\varepsilon(\overline{V}_C) + \eta(\overline{V}_C) \qquad (3\text{-}5)$$

$$\phi_2(\overline{V}_B, \overline{V}_C; P_B) = \theta_2(\overline{V}_B, P_B)\varepsilon(\overline{V}_C) + \eta(\overline{V}_C) \qquad (3\text{-}6)$$

Thus, $\phi_1 = \phi_2$ implies the validity of *Equation 3-4*; it also implies the validity of *Equation 3-3* because when $\phi_1 = \phi_2$ it is seen that $\theta_1(\overline{V}_A, P_A) = \theta_2(\overline{V}_B, P_B) = \theta$. We call θ the *empirical* temperature. Thus, for every system, it is possible to find a function θ of the mechanical variables (P and $\overline{V}$) such that when the systems are in thermal equilibrium with one another, they have the same value of θ. (Note that *Eq. 3-1*, which expresses the Zeroth Law, is essential to the argument.)

3.2   THE FIRST LAW—TRADITIONAL APPROACH

The traditional approach to the First Law is based on the assumption that the concepts of work and heat have already been established. The concept of work is taken from mechanics and will not be belabored here. The concept of heat, as alluded to before, is best described in terms of temperature changes. Now that temperature has been defined independently from the Zeroth Law, the concept of heat can be introduced in a straightforward manner by first defining heat capacity. If a system A, initially at temperature t_A, is brought in thermal contact with system B at temperature t_B at constant pressure and volume and the final temperature at equilibrium is t, then the ratio of the temperature changes, for infinitesimally small differences, $\delta t_A = t - t_A$ and $\delta t_B = t - t_B$ defines the ratio of the heat capacities of the two systems

$$(\delta t_A)/(\delta t_B) = -C_B/C_A \qquad (3\text{-}7)$$

Thus, if a particular value is assigned to one of the heat capacities, the other is then determined. The heat transfer for A is defined as

$$dq_A = C_A dt \quad \text{or} \quad q_A = \int C_A dt \qquad (3\text{-}8)$$

and for B

$$dq_B = C_B dt \quad \text{or} \quad q_B = \int C_B dt \qquad (3\text{-}9)$$

The First Law of Thermodynamics then reads (with our adopted sign convention for work)

$$\Delta E = q + w \quad \text{for macroscopic changes} \qquad (3\text{-}10)$$

and

$$dE = dq + dw \quad \text{for infinitesimal changes} \qquad (3\text{-}11)$$

where E is the *internal energy.*

The symbol ΔE stands for the energy difference between final and initial states of the system; that is, $\Delta E = E_B - E_A$. Thus, the first statement does not only imply that the internal energy is the sum of the heat and work energies, but also that E is a state function; that is, it is independent of the manner in which the state was obtained. The second statement implies that dE is an exact differential, despite the fact that dq and dw are generally not. (The distinction between exact and inexact differentials is sometimes denoted by crossing the d, or by writing a capital D. We will do neither, since dq and dw are the only inexact differentials used here, and we characterize work and heat by small letters, in contrast to the internal energy, and other thermodynamic functions to be introduced later, which we characterize by capital letters. Thus, dq and dw are inexact differentials and path dependent; dE is exact and path independent.)

The First Law can also be interpreted as a statement of conservation of energy.[1] Whatever the surrounding loses in the form of heat or work, the system gains in the form of internal energy and vice versa. There is no way to measure E itself, so this relation cannot be directly verified. What can be verified is that when a system, initially in a state A, changes to a state B, the quantity $q + w$ is path independent, that is, is independent of

---

[1]Since the advent of Einstein's Theory of Relativity, the Conservation Principle should refer not only to energy but to mass energy. In the absence of nuclear transformation, however, mass plays no role.

the manner in which the change is brought about. Also, when the system undergoes a cyclic change, A → B → A, then q = −w.

3.3 MATHEMATICAL INTERLUDE I: EXACT AND INEXACT DIFFERENTIALS

The notion of exact differentials plays such an important role in thermodynamics that it is of utmost importance to know how to manipulate them. Suppose we are given a differential expression of this form: $M(x, y)dx + N(x, y)dy$. Is this expression an exact differential? That is, can it be obtained from a function $f(x, y)$, which is a function of the same variables? If such a function exists, then

$$df = (\partial f/\partial x)_y \, dx + (\partial f/\partial y)_x \, dy \qquad (3\text{-}12a)$$

and

$$M(x, y) = (\partial f/\partial x)_y \qquad (3\text{-}12b)$$

$$N(x, y) = (\partial f/\partial y)_x \qquad (3\text{-}12c)$$

In other words, M and N are partial derivatives of $f(x, y)$. Thus, if two arbitrary functions $M(x, y)dx$ and $N(x, y)dy$ are combined, it is unlikely that *Equations 3-12b* and *3-12c* will be satisfied and the combination will unlikely be an exact differential.

A differential $df = M(x, y)dx + N(x, y)dy$ is exact, if any of the following statements are satisfied:

1) Its integral is path independent, i.e. $\int_A^B df = f_B - f_A$;
2) The integral along a closed contour is zero, i.e. $\oint df = 0$;
3) $(\partial M\,(x, y)/\partial y)_x = [\partial N(x, y)/\partial x]_y$.

Proof of *statement 3* is as follows:

$$df = Mdx + Ndy \text{ then} \qquad (3\text{-}13a)$$

$$\partial^2 f(x, y)/\partial y \partial x = [\partial/\partial y(\partial f/\partial x)_y]_x = (\partial M/\partial y)_x \qquad (3\text{-}13b)$$

$$\partial^2 f(x, y)/\partial x \partial y = [\partial/\partial x(\partial f/\partial y)_x]_y = (\partial N/\partial x)y \qquad (3\text{-}13c)$$

The left-hand sides of *Equations 3-13b* and *3-13c* are the same, since they only differ by the order of differentiation. Therefore, the right-hand sides of the equations must be equal.

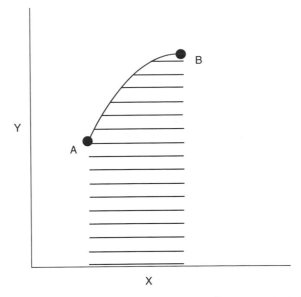

Figure 3.1a   Graphical representation of the integral $I = \int y \, dx$. The shaded area represents the value of the integral.

Proof of *statement 2* is mathematically somewhat complicated and will not be pursued here. *Statement 1* follows trivially from *statement 2*. If df is integrated from A to B along a particular path and returns from B to A along a different path, *statement 2* requires that the integrals be equal and opposite. In short, the integrals are independent of path and thus only initial and final states are important, as exemplified by *statement 1*.

Example: The differential $df = y\,dx$ is inexact since the integral depends on the path, as is obvious from the area under the curve in the diagram depicted in Figure 3.1a. The same holds true for the function $df = x\,dy$ (Figure 3.1b). However, the function $df = y\,dx + x\,dy$ is exact because the sum of the shaded areas (Figure 3.1c) is independent of the path—the integration limits only matter. These examples show how two differentials, which are intrinsically inexact, can add to give a function that is exact. The First Law of Thermodynamics is a case in point: dq and dw are, in general, not exact but dE is always exact.

3.4   THE FIRST LAW—AXIOMATIC APPROACH

In the *axiomatic* approach, *work* is assumed to be a well-defined mechanical concept but *heat* has yet to be defined. The goal is to define all thermal properties in terms of mechanical variables. This can be accomplished by

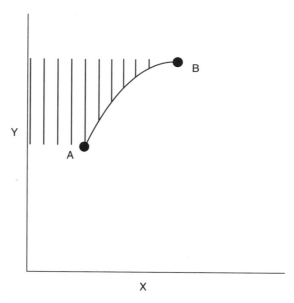

Figure 3.1b   Graphical representation of the integral $I = \int x \, dy$. The shaded area represents the value of the integral.

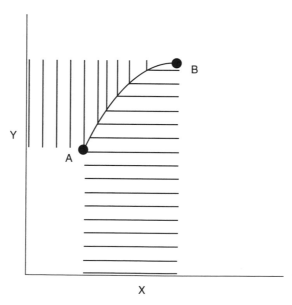

Figure 3.1c   Graphical representation of the integral $I = \int (y \, dx + x \, dy)$. The shaded area represents the value of the integral.

appealing to certain observations, such as the Joule (Paddle Wheel) experiment, which shows that if a system is brought from a state A to a state B adiabatically, *the work is always the same, regardless of path or source.*

Figure 3.2 is a schematic diagram of the Joule Paddle-Wheel experiment. An adiabatically encased fluid is stirred by a paddle wheel, which rotates as a result of placing weight on the tray or by some other mechanical device. The fluid is the system, and the paddle wheel is considered part of the surrounding. The rotating paddle wheel causes the temperature of the fluid to rise, thereby altering the state of the system. The change in the state is determined by observing the change in the temperature. It is found that the change in teperature, and thus the change in the state of the system, is *independent of the manner* in which the transition takes place. It is immaterial whether the transition is reversible or irreversible or whether it is produced by mechanical work or some other kind of work, such as electrical work.

Because the work in this adiabatic process is found to be independent of path, the differential dw_{ad} must be exact. Furthermore, because it is universally accepted (or believed) that energy—mass energy, since Einstein—cannot be created or destroyed, it is natural to assume that the lost adiabatic work, a form of energy, is transformed into another form of energy, the

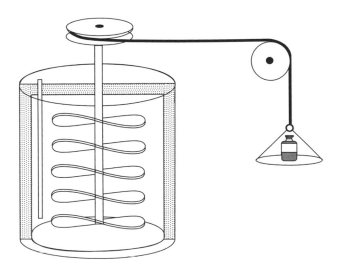

Figure 3.2 Joule's Paddle–Wheel Experiment. The weight in the pan sets the paddle wheel in motion, producing work on the fluid, which is dissipated as heat causing the temperature to rise.

internal energy, E. This energy is a property of the system, a state function, and must be independent of the manner in which it was created. These statements can be summarized concisely in the form of the First Law, which reads:

When a system makes a transition from state A to B by adiabatic means the change in internal energy is

$$\Delta E = E_B - E_A = w_{ad} \qquad (3\text{-}14a)$$

or

$$dE = dw_{ad} \qquad (3\text{-}14b)$$

Obviously, the differential dE must be exact, since dw_{ad} is exact and ΔE must be independent of the path.

What happens if the system is transformed from the same initial state A to the same final state B by work, w, which is nonadiabatic? There is no reason to assume that the *Equation 3-14a* will hold, and in fact

$$\Delta E = E_B - E_A \neq w \qquad (3\text{-}15)$$

To replace the inequality sign by an equal sign, a "correction" factor, q, must be added

$$\Delta E = q + w \qquad (3\text{-}16)$$

This correction factor defines the *heat*. Obviously, $q = w_{ad} - w$. [Actually, this equation does not really define the **concept** of heat, but rather the **measure** of heat.] In summary,

$$\Delta E = E_B - E_A = w_{ad} \qquad (3\text{-}17a)$$
$$\Delta E = E_B - E_A = q + w \qquad (3\text{-}18a)$$

or, in differential form

$$dE = dw_{ad} \qquad (3\text{-}17b)$$
$$dE = dq + dw \qquad (3\text{-}18b)$$

Note: Both ΔE and q are defined operationally and entirely in terms of the mechanical quantity work. Furthermore, the notion that the $E_B - E_A$ is path independent, and thus that E is a state function, is not a mere assumption but has an experimental basis.

Note: The axiomatic approach to the First Law of Thermodynamics tacitly assumes that it is always possible to reach an arbitrary state B from an arbitrary state A adiabatically. However, the Second Law of Thermodynamics (as we shall see later) shows this not to be the case. When it is not possible to reach B from A adiabatically, one should infer the energy change from the reverse process, namely, $\Delta E_{B \to A} = E_A - E_B$. The forward and reverse energy changes are equal and opposite. It will be shown later that if a forward process is not possible adiabatically, the reverse process *is* possible, although *irreversibly*. Measuring irreversible work functions may be very difficult. These difficulties are avoided in the traditional approach, which makes no use of adiabatic changes.

3.5 SOME APPLICATIONS OF THE FIRST LAW

The treatment here is restricted to reversible processes with no work other than pressure-volume work. Under these restrictions, $P_{ex} = P$, and

$$dq = dE - dw = dE + P_{ex}dV = dE + PdV \qquad (3\text{-}19)$$

3.5.1 Heat Capacity

The heat capacity is defined as

$$C = \lim_{(\delta T \to 0)} dq/\delta T = dq/dT \qquad (3\text{-}20)$$

In general, C is a function of temperature, mass, pressure, volume, and other variables. Usually, one or more of the variables are held constant, indicated here by a subscript on C. In particular,

$$C_V = dq_V/dT = (\partial E/\partial T)_V \qquad (3\text{-}21a)$$
$$C_P = dq_P/dT \qquad (3\text{-}21b)$$
$$C_{adiabatic} = 0 \qquad (3\text{-}21c)$$

3.5.2 Heat and Internal Energy

Let us regard E as a function of T and V, i.e., $E = E (T, V)$. Then,

$$dE = (\partial E/\partial V)_T dV + (\partial E/\partial T)_V dT \qquad (3\text{-}22)$$

and

$$dq = [(\partial E/\partial V)_T + P]dV + C_V dT \qquad (3\text{-}23)$$

For constant temperature $(T = T_A)$

$$q_T = \int_A^B (\partial E/\partial V)_T dV + \int_A^B PdV \tag{3-24}$$

$$= E(T_A, V_B) - E(T_A, V_A) + \int_A^B PdV \tag{3-25a}$$

$$= \Delta E + \int_A^B PdV \tag{3-25b}$$

For constant volume $(V = V_A)$

$$q_V = \int_A^B (\partial E/\partial T)_V = E(T_B, V_A) - E(T_A, V_A) \tag{3-26a}$$

$$= \Delta E \tag{3-26b}$$

For constant pressure $(P_A = P_B = P)$

$$q_P = \int_A^B dE + \int_A^B PdV = E(T_B, V_B) - E(T_A, V_A) + P\Delta V \tag{3-27a}$$

$$= \Delta E + P\Delta V \tag{3-27b}$$

Note: The symbol Δ is used extensively in chemical thermodynamics. Usually, the meaning is clear, but there are times when the notation is ambiguous. For example, one might conclude from *Equations 3-26b* and *3-27b* that $q_P = q_V + P\Delta V$, which is wrong. The ΔE values in the two equations are different, as is apparent when written in terms of the variables T_A, V_A, T_B, and V_B. The E values for the initial states are the same but not for the final states. Compare *Equations 3-25a, 3-26a, and 3-27a*.

3.5.3 Heat and Enthalpy

Equation 3-26b is particularly valuable because it enables one to calculate ΔE directly from measured heat changes at constant volume (q_V). Unfortunately, most chemical experiments are not done at constant volume but at constant pressure. Is there a thermodynamic function that relates simply to heat at constant pressure (q_p)? Yes, it is the *enthalpy.*
   The enthalpy is defined as

$$H = E + PV \tag{3-28}$$

H is a state function (path independent integral) because E and PV are state functions. If the system is taken from state A to state B,

$$\Delta H = \Delta E + \Delta(PV) = E(T_B, V_B) - E(T_A, V_A) + (P_B V_B - P_A V_A)$$

$$(3\text{-}29)$$

At constant pressure, $P_A = P_B = P$, the right-hand side of the equation is identical to *Equation 3-27a*; thus

$$q_P = \Delta H \qquad\qquad (3\text{-}30)$$

Another way to reach the same conclusion is to observe that at constant pressure

$$\Delta H = \Delta E + \Delta(PV) = q + w + P\Delta V = q_P - P\Delta V + P\Delta V = q_P \quad (3\text{-}30a)$$

Writing H as a function of T and P shows that

$$dH = dE + PdV + VdP \qquad\qquad (3\text{-}31)$$

$$= dq - PdV + PdV + VdP = dq + VdP \qquad (3\text{-}32)$$

or

$$dq = dH - VdP \qquad\qquad (3\text{-}33)$$

Thus, at constant P

$$C_P = dq_P/dT = (\partial H/\partial T)_P \qquad\qquad (3\text{-}34)$$

Similarly, from the expression

$$dq = dE + PdV \qquad\qquad (3\text{-}35)$$

we obtain for constant volume

$$C_V = dq_v/dT = (\partial E/\partial T)_V \qquad\qquad (3\text{-}36)$$

Finally, the reader is reminded that the relations between q_v and ΔE and between q_P and ΔH are not general; rather, the relations are predicated on the assumptions that there is no work other than PV work and that the process is reversible.

3.6 MATHEMATICAL INTERLUDE II:
PARTIAL DERIVATIVES

The partial derivative relations that are used most frequently for the purposes of the course outlined in this book are derived here. They are of two types: (1) relations between partials of dependent variables and (2) relations between partials with different subscripts.

3.6.1 Relations Between Partials of Dependent Variables

Consider three interrelated variables x, y, and z. If z is treated as a function of x and y, that is $z = z(x, y)$, then

$$dz = (\partial z/\partial x)_y dx + (\partial z/\partial y)_x dy \tag{3-37}$$

However, if y is taken to be a function of x and z: $y = y(x, z)$, then

$$dy = (\partial y/\partial x)_z dx + (\partial y/\partial z)_x dz \tag{3-38}$$

Substituting *Eq. 3-38* in *Eq. 3-37* gives

$$dz = [(\partial z/\partial x)_y + (\partial z/\partial y)_x(\partial y/\partial x)_z]dx + (\partial z/\partial y)_x(\partial y/\partial z)_x dz \tag{3-39}$$

Equating the coefficients of the differentials yields

$$1 = (\partial z/\partial y)_x(\partial y/\partial z)_x \tag{3-40a}$$

or

$$(\partial z/\partial y)_x = 1/(\partial y/\partial z)_x \tag{3-40b}$$

and

$$(\partial z/\partial x)_y + (\partial z/\partial y)_x(\partial y/\partial x)_z = 0 \tag{3-41a}$$

or

$$(\partial y/\partial x)_z = -(\partial z/\partial x)_y/(\partial z/\partial y)_x \tag{3-41b}$$

or

$$(\partial z/\partial x)_y(\partial x/\partial y)_z(\partial y/\partial z)_x = -1 \tag{3-41c}$$

3.6.2   Relations Between Partials with Different Subscripts

Consider a function Γ of x and y, i.e. $\Gamma = \Gamma(x,y)$. The differential is

$$d\Gamma = (\partial\Gamma/\partial x)_y dx + (\partial\Gamma/\partial y)_x dy \qquad (3\text{-}42a)$$

But Γ can be regarded also a function of z and y. Thus, differentiating $d\Gamma$ in *Equation 3-42a* with respect of y, holding z constant, gives

$$(\partial\Gamma/\partial y)_z = (\partial\Gamma/\partial y)_x + (\partial\Gamma/\partial x)_y(\partial x/\partial y)_z \qquad (3\text{-}42b)$$

The latter equation is particularly useful in relating partials such as $(\partial E/\partial T)_P$ to $(\partial E/\partial T)_V$ etc.

3.7   OTHER APPLICATIONS OF THE FIRST LAW

The following applications are intimately based on the mathematical techniques outlined in Section 3.6 and are useful exercises of those techniques.

3.7.1 $C_P - C_V$

In addition to mathematical formulas developed in the previous section, we shall also make use of the so-called *thermodynamic equations of state*, to be derived later, namely

$$(\partial E/\partial V)_T = T(\partial P/\partial T)_V - P \qquad (3\text{-}43)$$

$$(\partial H/\partial P)_T = V - T(\partial V/\partial T)_P \qquad (3\text{-}44)$$

From the definitions of C_P and C_V in *Equations 3-34* and *3-36*, we obtain

$$C_P - C_V = (\partial H/\partial T)_P - (\partial E/\partial T)_V$$

$$= (\partial E/\partial T)_P + P(\partial V/\partial T)_P - (\partial E/\partial T)_V \qquad (3\text{-}45a)$$

$$= (\partial E/\partial T)_V + [(\partial E/\partial V)_T(\partial V/\partial T)_P] + P(\partial V/\partial T)_P$$

$$- (\partial E/\partial T)_V \qquad (3\text{-}45b)$$

$$= [(\partial E/\partial V)_T + P](\partial V/\partial T)_P \qquad (3\text{-}45c)$$

and using *Equation 3-43*, gives

$$C_P - C_V = T(\partial P/\partial T)_V(\partial V/\partial T)_P \qquad (3\text{-}46)$$

Finally, using the coefficient of thermal expansion,

$$\alpha = 1/V(\partial V/\partial T)_P \qquad (3\text{-}47a)$$

and the compressibility,

$$\kappa = -1/V(\partial V/\partial P)_T \qquad (3\text{-}47b)$$

yields

$$C_P - C_V = TV\,\alpha^2/\kappa \qquad (3\text{-}48)$$

Eqs. 3-46 and *3-48* are general. For 1 mol of an ideal gas, $P\overline{V} = RT$, and

$$\overline{C}_P - \overline{C}_V = R \qquad (3\text{-}49)$$

3.7.2   Isothermal Change, Ideal Gas (1 mol)

$$(\partial E/\partial V)_T = T(\partial P/\partial T)_V - P = TR/\overline{V} - P = 0 \qquad (3\text{-}50a)$$

At constant temperature, $\Delta E = 0$, and

$$q = -w = \int_1^2 PdV = RT \int_1^2 d\overline{V}/\overline{V}$$

$$= RT\ln \overline{V}_2/\overline{V}_1 = RT\ln P_1/P_2 \qquad (3\text{-}50b)$$

3.7.3   Adiabatic Change, Ideal Gas (1 mol)

$$dq = dE - dw = \overline{C}_V\,dT + Pd\overline{V} = 0 \qquad (3\text{-}51a)$$

$$\overline{C}_V\,dT/T + R\,d\overline{V}/\overline{V} = \overline{C}_V\,d\ln T + Rd\ln \overline{V} = 0 \qquad (3\text{-}51b)$$

Using *Eq. 3-49* and replacing $\overline{C}_P/\overline{C}_V$ by γ, we get, after integration

$$\ln T_2/T_1 + (\gamma - 1)\ln \overline{V}_2/\overline{V}_1 = 0 \qquad (3\text{-}52a)$$

or

$$(T_2/T_1)(\overline{V}_2/\overline{V}_1)^{\gamma-1} = 1 \qquad (3\text{-}52b)$$

If we replace T_2/T_1 by $P_2\overline{V}_2/P_1\overline{V}_1$, we get

$$(P_2/P_1)(\overline{V}_2/\overline{V}_1)^\gamma = 1 \qquad (3\text{-}53a)$$

Equations 3-52b and *3-53a* are often expressed in the form

$$T\overline{V}^{\gamma-1} = \text{const} \quad \text{and} \quad P\overline{V}^\gamma = \text{const} \qquad (3\text{-}53b)$$

Another way to obtain *Equation 3-53a* is to derive it directly from *Equation 3-54*, which expresses dq in terms of both C_P and C_V

$$dq = C_V(\partial T/\partial P)_V \, dP + C_P(\partial T/\partial V)_P \, dV \qquad (3\text{-}54)$$

To prove this relation, first consider E to be a function of V and P

$$\begin{aligned} dq &= dE + PdV \\ &= (\partial E/\partial V)_P \, dV + PdV + (\partial E/\partial P)_V \, dP \\ &= (\partial H/\partial V)_P \, dV + (\partial E/\partial P)_V \, dP \\ &= (\partial H/\partial T)_P(\partial T/\partial V)_P \, dV + (\partial E/\partial T)_V(\partial T/\partial P)_V \, dP \\ &= C_P(\partial T/\partial V)_P dV + C_V(\partial T/\partial P)_V \, dP \end{aligned}$$

For 1 mol of an ideal gas $(\partial T/\partial \overline{V})_P = P/R$ and $(\partial T/\partial P)_V = \overline{V}/R$ and in an adiabatic transition

$$P\,\overline{C}_P d\overline{V} + \overline{V}\,\overline{C}_V dP = 0 \qquad (3\text{-}55)$$

yielding

$$(1/\overline{V})(\overline{C}_P/\overline{C}_V)dV + dP/P = \gamma d\ln\overline{V} + d\ln P = 0 \qquad (3\text{-}56)$$

$$(\overline{V}_2/\overline{V}_1)^\gamma(P_2/P_1) = 1 \text{ or } P\overline{V}^\gamma = \text{const} \qquad (3\text{-}57)$$

which are the same as *Equations 3-53a* and *3-53b*.

3.7.4 The Joule and the Joule-Thomson Coefficients

The *Joule Coefficient* is defined as

$$\mu_J = (\partial T/\partial V)_E \qquad (3\text{-}58)$$

Obviously,

$$\mu_J = -(\partial E/\partial V)_T/(\partial E/\partial T)_V = -(\partial E/\partial V)_T/C_V \qquad (3\text{-}59)$$

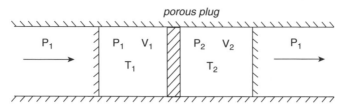

Figure 3.3   The Joule-Thomson Experiment. A gas at constant P_1, V_1, and T_1 in the adiabatic container on the left is forced through porous plug to the adiabatic chamber on the right, maintained at constant P_2, V_2, and T_2. The entire device is thermally insulated.

Joule measured the change in temperature when a gas, adiabatically enclosed, is allowed to expand into a vacuum. Because, in this experiment, both w (expansion into a vacuum) and q (adiabatic expansion) are zero, so must be ΔE; hence, the constancy of E in *Eq. 3-58*. Joule found that there was no noticeable change in T in this expansion and concluded (e.g., see *Eq. 3-59*) that $(\partial E/\partial V)_E = 0$, in other words that the energy of the gas was independent of the volume. This is approximately true for very dilute gases; it is exact for ideal gases but not for real gases. For example, it will be shown later that for a van der Waals gas, $(\partial E/\partial \overline{V})_T = a/\overline{V}^2$, where a is one of the van der Waals constants, and thus the Joule coefficient cannot strictly be zero. Joule's experiment was rather crude.

The *Joule-Thomson Coefficient* is defined as

$$\mu_{JT} = (\partial T/\partial P)_H \qquad (3\text{-}60)$$

which yields

$$\mu_{JT} = -(\partial H/\partial P)_T/(\partial H/\partial T)_P = -(\partial H/\partial P)_T/C_P \qquad (3\text{-}61)$$

In this experiment, a gas is slowly forced through a porous plug from chamber 1 to chamber 2. The pressures of chamber 1 and chamber 2 are, respectively, P_1 and P_2 and are kept constant during the operation. The volume of chamber 1 changes from V_1 to 0 and the volume of chamber 2 changes from 0 to V_2 (see Figure 3.3). The system is adiabatically enclosed. The overall change in this experiment is

$$\Delta E = E_2 - E_1 = w = -\int_1^2 PdV = P_1V_1 - P_2V_2 \qquad (3\text{-}62)$$

Thus,

$$E_1 + P_1V_1 = E_2 + P_2V_2 \qquad (3\text{-}63)$$

or

$$H_1 = H_2 \qquad (3\text{-}64)$$

showing that H is constant in the process, as indicated in the definition of the Joule-Thomson coefficient. In the Joule-Thomson experiment, the temperature is observed to change with pressure and has been used to liquefy gases.

Note: The Joule-Thomson procedure does not always result in cooling. For cooling to occur, $(\partial H/\partial P)_T$ has to be negative (*why?*); for most gases at ordinary temperatures it is. However, there are exceptions: H_2 and He heat up at room temperature. With the use of the virial form of the equation of state (Chapter 8), it is easy to show that $(\partial H/\partial P)_T = b - 2a/RT$ for a van der Waals gas, and obviously this derivative can be positive or negative, depending on the temperature and the van der Waals constants. The temperature for which the partial $(\partial H/\partial P)_T$ is zero is called the inversion temperature (T_i). Thus, $(\partial H/\partial P)_{T_i} = 0$.

EXERCISES

1. The van der Waals constants a and b for N_2 are, respectively, 1.390 $L^2 \cdot$atm and 0.03913 L/mol. Calculate the inversion temperature (T_i).
2. Repeat the calculation for He. The van der Waals constants a and b are respectively 0.03142 $L^2 \cdot$atm and 0.02370 L/mol.

CHAPTER 4

THE LAWS OF THERMODYNAMICS II

4.1 THE SECOND LAW—TRADITIONAL APPROACH

Some processes occur spontaneously, others do not. For example, if system A is hotter than B and A and B are in thermal contact, heat will flow from A to B. The opposite—heat flowing from B to A—does not occur, at least not spontaneously, although that would not violate the First Law. In chemical reactions, certain reactions proceed spontaneously, whereas others do not. In general, reactions will go spontaneously if the sum of the enthalpies of the products is less than the sum of the enthalpies of the reactants, i.e., if ΔH_{rec} is negative (exothermic!). However, there are exceptions. Liquids will generally mix when the enthalpy of mixing (ΔH_{mix}) is negative but there are liquids (benzene and toluene, for example) that mix readily when $\Delta H_{mix} > 0$. Obviously, the requirement that the final enthalpy be less than the initial is not sufficient to serve as a criterion for spontaneity. The same can be said about the internal energy (E). It turns out that many physical phenomena cannot be explained on the basis of the First Law alone; thus a new law—the Second Law—is needed.

In Chapter 4, we will first discuss the Second Law based on the Traditional Approach. We will then present a nonrigorous treatment of the Axiomatic Approach introduced by Carathéory. The usual development of the

Thermodynamics and Introductory Statistical Mechanics, by Bruno Linder
ISBN 0-471-47459-2 © 2004 John Wiley & Sons, Inc.

traditional approach starts with statements (called Principles) by Clausius or by Kelvin-Planck.

The *Clausius Principle* states that it is not possible to devise an engine (operating in a cycle) that has the sole effect of transferring heat from a colder body to a hotter body (i.e., without at the same time converting a certain amount of heat into work).

The *Kelvin-Planck Principle* states that it is not possible to devise an engine (operating in a cycle) that has the sole effect of extracting heat from a heat reservoir and converting it all into work (i.e., without at the same time transferring heat from a hotter to colder body).

It can be shown that, as a consequence of one or the other of these principles, a system must posses a property, which is a state function—called the entropy—whose differential is related to the reversible heat exchanged divided by the temperature, namely $dS = dq_{rev}/T$. Another consequence is that the entropy of an isolated system can never decrease.

The standard way of proving those statements is by means of Carnot cycles. The *Carnot cycle* is a reversible cycle consisting of two (reversible) isotherms and two (reversible) adiabats. There are two heat reservoirs, one at the empirical temperature θ_2 and one at the temperature θ_1. Basically, a gas is allowed to expand isothermally and reversibly at the higher temperature θ_2, followed by an adiabatic expansion to the lower temperature θ_1. This is followed by an isothermal compression at θ_1 and finally by an adiabatic compression to θ_2 (see Figure 4.1). Heat is extracted from the reservoir at the higher temperature, θ_2, and released at the lower temperature, θ_1. During this process, work is done on the surroundings.

It is convenient to represent the heat engine as in the diagram in Figure 4.2. The quantities q_2 and q_1 represent the heats exchanged

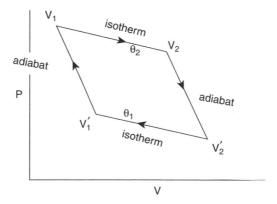

Figure 4.1 Carnot Cycle. The curves represent two isotherms and two adiabats. The upper isotherm is at a higher temperature.

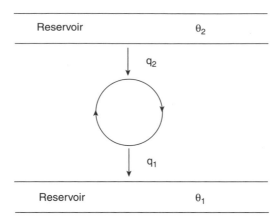

Figure 4.2a Schematic representation of a Carnot cycle running in a clockwise direction and transferring heat from the high temperature reservoir to the lower one.

respectively at θ_2 and θ_1, and q may have either a positive or a negative value, depending on whether heat is absorbed or emitted. (In a heat engine, q_2 will be positive and q_1 negative.) The arrow in Figure 4.2a serves to denote that the cycle runs in a forward direction, as the Carnot cycle depicted in Figure 4.1. The efficiency of an engine is defined as the work done by the engine in a complete cycle divided by the heat absorbed at the higher temperature reservoir; that is

$$\xi = -w/q_2 \qquad (4\text{-}1)$$

The work done by the system on the surroundings must be represented as $-w$, because, by our convention, w represents the work done on the system. Note that, in the cyclic process, $\Delta E = q + w = 0$, since the system returns to its origin and E is a state function. Accordingly, the overall heat $q = q_1 + q_2 = -w$ must be positive, since $-w$ represents the work done on the surroundings and thus is positive. Replacing $-w$ by $q_1 + q_2$ in *Equation 4-1* yields

$$\xi = (q_1 + q_2)/q_2 = 1 + q_1/q_2 \qquad (4\text{-}2)$$

To proceed, we make use of a theorem, from Carnot, that is essential to the traditional development of the Second Law of Thermodynamics. *Carnot's Theorem states that Carnot cycles operating reversibly between the same temperatures have the same efficiency.*

To prove this, we assume that there is an engine operating between the temperatures θ_2 and θ_1 which has an efficiency ξ^* greater than ξ. We show

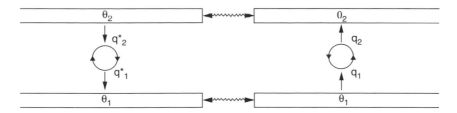

Figure 4.2b Schematic representation of two coupled Carnot engines. The starred cycle runs in the forward direction (clockwise); the unstarred runs in the reverse direction (counter-clockwise).

that this assumption violates both the Clausius and the Kelvin-Planck Principles.

First let us consider the Clausius Principle. The assumption $\xi^* > \xi$ leads to the supposition that

$$-w^*/q_2^* > -w/q_2 \qquad (4\text{-}3)$$

Let us, for simplicity, adjust the two engines in such a way that $-w^* = -w$. Then, $1/q_2^* > 1/q_2$ and therefore $1/q_1^* < 1/q_1$. We now couple the two engines (see Figure 4.2b), running the starred-engine in the forward direction (clockwise) and the nonstarred engine in the reverse direction (counter-clockwise). The heat changes in the reverse direction are opposite to what they would have been in the forward direction. Thus, the heat changes of the reverse engine are $-q_2$ and $-q_1$. We conclude that at the high temperature the overall heat transferred is $q_2^* - q_2 < 0$, i.e., heat is released and at low temperature $q_1^* - q_1 > 0$, i.e., heat is absorbed, in violation of the Clausius Principle.

If we use the Kelvin-Planck Principle and again assume that $\xi^* > \xi$ or that $\frac{-w^*}{q_2^*} > -\frac{w}{q_2}$ and adjust the systems so that $q_2^* = q_2$, we find that $-w^* > -w$ or $q_2^* + q_1^* > q_2 + q_1$ so that $q_1^* > q_1$. Again, if we run the starred engine in a forward direction and the nonstarred one in the reverse direction (where at θ_1 the heat evolved is $-q_1$), we find that at the high temperature $q_2^* - q_2 = 0$ and at the low temperature $q_1^* - q_1 > 0$. The overall work is $w^* - w < 0$. The net result is that heat is converted entirely into work without the transfer of heat from the hotter to the colder reservoir—in violation of the Kelvin-Planck Principle. Thus, both the Clausius Principle and the Kelvin-Planck Principle establish the validity of Carnot's Theorem.

The next obvious step is to calculate the efficiency of a particular system. If we know it for one system, then we know it for all. The easiest thing to do is to determine the efficiency of an ideal gas, as is done typically in elementary physical chemistry (and summarized in Section 4.2.1). The ideal gas law is used to relate the engine efficiency to the absolute temperature. However, this is not necessary. It is possible to obtain the absolute temperature directly from the heat relations of coupled engines (Section 4.2.2) and thereby avoid the use of the ideal gas or any other system for that matter.

4.2 ENGINE EFFICIENCY: ABSOLUTE TEMPERATURE

4.2.1 Ideal Gas

In elementary treatments, the temperature is usually taken to be defined by the ideal gas law. Referring to Figure 4.1 and replacing the reservoir temperatures θ_2 and θ_1 by T_2 and T_1, we obtain for the sequence $(V_1, T_2) \rightarrow (V_2, T_2) \rightarrow (V_2', T_1) \rightarrow (V_1', T_1) \rightarrow (V_1, T_2)$

$$q_2 + q_1 = RT_2 \ln \overline{V}_2/\overline{V}_1 + RT_1 \ln \overline{V}_1'/\overline{V}_2' \tag{4-4}$$

which, by applying (Eq. 3-52b) to the adiabats gives $V_2/V_2' = V_1/V_1'$ or $V_2/V_1 = V_2'/V_1'$ and so

$$\varepsilon = -w/q_2 = 1 + q_1/q_2$$
$$= [RT_2 \ln(V_2/V_1) - RT_1 \ln(V_2/V_1)]/RT_2 \ln(V_2/V_1) \tag{4-5}$$
$$= 1 - T_1/T_2 \tag{4-6}$$

which gives

$$q_1/T_1 + q_2/T_2 = 0 \tag{4-7}$$

4.2.2 Coupled Cycles

The extremely simple derivation shown in Section 4.2.1 is based on the properties of an ideal gas. Avoiding the use of an ideal gas (or any other system), we must first define the absolute temperature. This is done by considering two coupled Carnot cycles arranged as in the diagram of Figure 4.3. The empirical temperatures of the heat reservoirs are in the order $\theta_3 > \theta_2 > \theta_1$. Engine A absorbs from the reservoir at θ_3 an amount of heat, q_3, and emits an amount q_2^* to reservoir θ_2. Engine B absorbs an amount of heat q_2 at θ_2 and emits an amount q_1 at θ_1. Figure 4.3 represents $q_2^* = -q_2$.

The efficiency of A, $\xi_A = 1 + q_2^*/q_3$, depends only on θ_2 and θ_3 and so

$$-q_2^*/q_3 = 1 - \xi_A = f(\theta_2, \theta_3) \tag{4-8}$$

Similarly, the efficiency of B, $\xi_B = 1 + q_1/q_2$ and so

$$-q_1/q_2 = 1 - \xi_B = f(\theta_1, \theta_2) \tag{4-9}$$

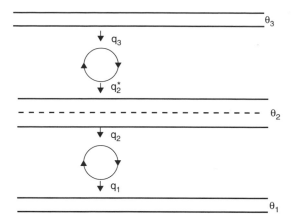

Figure 4.3 Schematic representation of two coupled Carnot cycles transferring heat from reservoir at θ_3 to θ_2 and from θ_2 to θ_1.

If we consider the coupled engine as a single unit, operating between θ_3 and θ_1, the efficiency would be $\xi_{AB} = 1 - q_1/q_3$ and thus

$$-q_1/q_3 = 1 - \xi_{AB} = f(\theta_1, \theta_3) \qquad (4\text{-}10)$$

Accordingly, we can write

$$-q_1/q_3 = (-q_1/q_2)(q_2/q_3) = (-q_1/q_2)(-q_2^*/q_3)$$
$$= f(\theta_1, \theta_2)f(\theta_2, \theta_3) \qquad (4\text{-}11)$$

and, using *Equations 4-9* and *4-10*, we can obtain

$$f(\theta_1, \theta_2) = f(\theta_1, \theta_3)/f(\theta_2, \theta_3) \qquad (4\text{-}12)$$

Because the left-hand side of *Equation 4-12* is independent of θ_3 but is always equal to the right-hand side, which depends on θ_3, the function $f(\theta_1, \theta_2)$ must be of the form

$$f(\theta_1, \theta_2) = T(\theta_1)/T(\theta_2) \qquad (4\text{-}13)$$

where T is a universal function of the empirical temperature (θ) and independent of the substance. We define $T(\theta_2) = T_2$, etc., and call it the absolute temperature. Thus, from *Eqs. 4-9* and *4-13*, we obtain

$$-q_1/q_2 = T_1/T_2 \qquad (4\text{-}14)$$

and, operationally, the ratio T_1/T_2 can be obtained from $-q_1/q_2$. To determine the absolute scale unambiguously, Kelvin suggested taking the difference between the boiling point of water [$T_{bp}(H_2O)$] and the freezing point of water [$T_{fp}(H_2O)$] and setting it equal to 100. This gives the Kelvin scale the same size as the Celsius scale.

Finally, rearranging *Equation 4-14* gives

$$q_2/T_2 + q_1/T_1 = 0 \qquad (4\text{-}15)$$

which is the same as *Equation 4-7* but is obtained here without recourse to an ideal gas or any other substance. *Equation 4-15* is fundamental to (reversible) Carnot cycles and essential for further development of the subject.

4.3   GENERALIZATION: ARBITRARY CYCLE

The above result obtained for a reversible Carnot cycle can be generalized to an arbitrary cycle. To accomplish this, we suppose that, when the system traverses the cyclic path, it exchanges heat with a series of reservoirs at temperatures T_1, T_2, etc. We replace the reversible cycle by a sum of small Carnot cycles (as shown in Figure 4.4), each of which operates quasi-statically. It is seen that, when all the small Carnot cycles are completed, each adiabat has been traversed twice, once in the forward direction and once in the reverse direction, effectively cancelling each other. What remains is the heat exchange of the outer path. Denoting the heat exchanges by the small cycles δq_i, the net result for the arbitrary reversible path is

$$\Sigma_i \delta q_i/T = 0 \qquad (4\text{-}16)$$

and, in the limit as $\delta q_i \to 0$

$$\oint dq_{rev}/T = 0 \qquad (4\text{-}17)$$

This clearly shows that dq_{rev}/T is an exact differential.

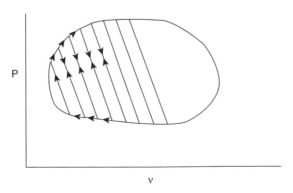

Figure 4.4   Schematic representation of an arbitrary reversible cycle as a sum of Carnot cycles.

4.4 THE CLAUSIUS INEQUALITY

Let us return to the discussion of the efficiency of the Carnot cycle. We saw
that, if we coupled the (assumed) more efficient starred (*) cycle to the lesser
efficient one and ran the starred cycle in a forward direction and the
unstarred in a reverse direction, we arrived at a contradiction of the Clausius
Principle. We concluded that the supposition that the starred cycle is more
efficient than the unstarred one is wrong and that ξ^* cannot be greater than ξ.
But can it be less than ξ? Not if the starred cycle is reversible. If so, we
could run the unstarred cycle in a forward direction and the starred cycle
in a reverse direction and arrive at the same contradiction. Therefore, if
both engines are capable of performing in forward and reverse directions,
i.e., if both engines are *reversible*, they must have the same efficiency.
But what if the starred engine is *irreversible*? (The unstarred engine runs
a Carnot cycle and is always reversible!) Then, the starred engine cannot
be coupled to the unstarred one in a reverse direction, and the efficiency
ξ^* cannot be greater than or equal to that of the reversible engine: In short,
ξ^* must be less than ξ.

In summary, because $\xi = 1 - T_1/T_2$ it follows that $\xi^* < 1 - T_1/T_2$ and
thus for the irreversible case

$$1 + q_1/q_2 < 1 - T_1/T_2$$

or

$$q_1/T_1 + q_2/T_2 < 0 \qquad (4\text{-}18)$$

For an arbitrary irreversible cycle, taken along a closed contour,
$\oint dq_{irrev}/T < 0$. We may combine the results for the reversible and irreversible cycles by writing

$$\oint dq/T \leq 0 \qquad (4\text{-}19)$$

where the $=$ sign refers to a reversible cycle and the $<$ sign to an irreversible
cycle. *Equation 4-19* is known as the Claudius Inequality. As stated earlier,
because $\oint dq_{rev}/T = 0$, the differential dq_{rev}/T is an exact differential, sug-
gesting the existence of a property of the system, which is a state function,
called the entropy and denoted as S, whose differential is related to the
element of reversible heat

$$dS = dq_{rev}/T \qquad (4\text{-}20a)$$

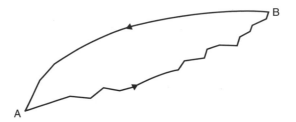

Figure 4.5 Schematic diagram of a cyclic process, representing a reversible path from B to A (smooth curve), and an irreversible path (zigzag curve) from A to B.

Also, the integral between two states is path independent

$$\Delta S = \int_A^B dq_{rev}/T = S_B - S_A \qquad (4\text{-}20b)$$

If dq is not reversible, we can construct a cyclic process, which proceeds from A to B irreversibly and returns from B to A by a reversible path (see Figure 4.5). Because part of the cycle is irreversible, the overall cycle is irreversible and thus

$$\oint dq/T = \int_A^B dq_{irr}/T + \int_B^A dq_{rev}/T < 0 \qquad (4\text{-}21a)$$

or

$$\int_A^B dq_{irr}/T < - \int_B^A dq_{rev}/T \qquad (4\text{-}21b)$$

or

$$\int_A^B dq_{irr}/T < \int_A^B dq_{rev}/T \qquad (4\text{-}21c)$$

We conclude that

$$\Delta S = S_B - S_A > \int_A^B dq_{irr}/T \qquad (4\text{-}22)$$

For an isolated (adiabatic) system undergoing an irreversible change, $dq_{irr} = 0$ and $\Delta S_{iso} > 0$. In general, for an isolated system (reversible or irreversible), we can write

$$\Delta S_{\text{isolated system}} \geq 0 \qquad (4\text{-}23)$$

where the $=$ sign refers to the reversible transformation and the $>$ sign to the irreversible one.

This is an important conclusion. It says that the entropy of an isolated system can never decrease: if the system undergoes a reversible change, its entropy stays the same; if the change is irreversible, the entropy must increase. Because all spontaneous or naturally occurring processes are irreversible (as mentioned before), this statement provides a clue on how to distinguish between a process in equilibrium and a spontaneous process.

4.5 THE SECOND LAW—AXIOMATIC APPROACH (CARATHÉODORY)

Traditional approaches to the Second Law of Thermodynamics based on the Clausius or the Kelvin-Planck Principle have been criticized by a some scholars [for example, see Max Born, (1921)] that such important concepts as entropy should come about as an addendum to a discussion of heat engines. Some individuals objected to the use of reversible Carnot cycles, because of the implication of frictionless pistons, infinite heat reservoirs, and so forth. Although these are part of the environment and not the system, to the mathematician. Carathéodory arguments based on Carnot cycles were unsatisfactory.

Here, we give a brief sketch of Carathéodory's Axiomatic approach to the Second Law. This approach is based on two statements: "Carathéodory's Principle" and "Carathéodory's Theorem." Carathéodory's Principle is a statement that must be accepted without proof, like the Clausius or Kelvin-Planck Principle. Carathéodory's Theorem is a mathematical theorem that can be proved (and was proved by Carathéodory!). The theorem is mathematically involved and will not be developed here. Rather, we will discuss the theorem in sufficient detail to highlight the arguments that lead to the concept and properties of entropy.

Carathéodory's Principle states the following: *in the neighborhood of any given state of a thermodynamic system, there exist states that cannot be reached from it by any quasi-static adiabatic process.* (This is also called the Principle of Adiabatic Inaccessibility.)

It may seem strange, but this principle can be shown to be consistent with predictions based on the Clausius Principle or the Kelvin-Planck Principle. To show this, consider a system with states whose generalized coordinates are θ (the empirical temperature) and X (an extensive coordinate, such as the volume, V). Figure 4.6 depicts two (reversible) isotherms and two (reversible) adiabats. It is assumed (consistent with $X = V$) that in the transition $1' \rightarrow 1$ heat is absorbed. We assert that $2'$ cannot be reached from 1 by

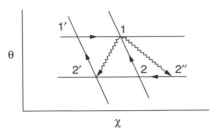

Figure 4.6   Illustration showing that points to the left of the adiabat 1-2 cannot be reached from the adiabat in accordance with the traditional approach to the Second Law; points to the right of the adiabat can be reached but irreversibly.

an adiabatic process. To show this, assume that $2'$ is accessible adiabatically from 1. Then, in the cycle $1 \rightarrow 2' \rightarrow 1' \rightarrow 1$, $\Delta E = 0$ and $q = q_{1 \rightarrow 2'} + q_{2' \rightarrow 1'} + q_{1' \rightarrow 1} = 0 + 0 + q_{1' \rightarrow 1} > 0$. Therefore, w must be negative. In other words, heat is converted entirely into work without other changes, in violation of the Kelvin-Planck statement. Thus, the supposition that B can be reached from A adiabatically is false. In fact, all points to the left of the adiabat through 1 are inaccessible adiabatically from 1. Points to the right of the adiabat $1 \rightarrow 2$ can be reached adiabatically, but the process must be irreversible. Consider cycle $1 \rightarrow 2'' \rightarrow 2 \rightarrow 1$. Here, $q = 0 + q_{2'' \rightarrow 2} + 0 < 0$ and therefore, $w > 0$, indicating that work is done on the system and converted entirely into heat, which is no violation of the Kelvin-Planck Principle. However, the path $1 \rightarrow 2''$ must be irreversible. If the path were reversible, the steps could be retraced along the path $1 \rightarrow 2 \rightarrow 2'' \rightarrow 1$, which will produce the inequalities $q > 0$ and $w < 0$ in violation of Kelvin-Planck. The conclusions reached so far apply equally well to systems of more than two variables. If a system is dependent on θ and on more variables than X, the adiabatic lines will have to be replaced by multi-dimensional surfaces parallel to the θ-axis. The inaccessible region will be a multi-dimensional volume, etc. The region to the left of the adiabatic surface will be inaccessible adiabatically from a point on the surface and the region to the right of the surface will be accessible but irreversibly.

Recall that in the discussion of the Axiomatic Approach to the First Law (Section 3-4), the question was raised of whether it is always possible to reach a state B from A adiabatically. Obviously, in light of the present discussion, the answer is no! But if B cannot be reached adiabatically from A, it can be shown that A can be reached adiabatically from B, but only irreversibly. Even so, by obtaining the work change from B to A (and thus the energy change), we obtain in effect the energy change from A to B. Irreversibility does not come into play in the First Law.

Note: It is obvious that two (reversible) adiabats cannot intersect. Once Carathéodory's Principle is accepted, the concept of entropy can be defined without recourse
to Carnot cycles by using mathematical properties of certain linear differential
equations called Pfaffian expressions, summarized in the following mathematical
interlude.

4.6 MATHEMATICAL INTERLUDE III: PFAFFIAN DIFFERENTIAL FORMS

A differential expression of the form

$$dL = \Sigma_i X_i \, dx_i \qquad (4\text{-}24)$$

where X_i is a function of the variables $x_1, x_2, \ldots, x_r$, is called a Pfaffian
differential form. The equation

$$dL = \Sigma_i X_i \, dx_i = 0 \qquad (4\text{-}25)$$

is called Pfaffian differential equation. The differential forms are sometimes
exact, although generally they are not.

We consider three special cases.

1) dL is exact. Then

$$\partial X_i / \partial x_j = \partial X_j \, \partial x_i \quad \text{for all } i, j = 1, 2, \ldots, r \qquad (4\text{-}26)$$

2) dL in not exact, but has an integrating factor.

An integrating factor, $\Lambda(x_1, \ldots, x_r)$, is a factor that turns the inexact
differential dL into an exact one, which we represent as dσ. Writing

$$\Lambda dL = d\sigma = \Sigma_i (\Lambda X_i) \, dx_i = \Sigma_i (\partial \sigma / \partial x_i)_{j \neq i} \, dx_i \qquad (4\text{-}27)$$

The reciprocity relations require that

$$\partial(\Lambda X_i) / \partial x_j = \partial(\Lambda X_j) / \partial x_i \qquad (4\text{-}28)$$

Note: If a Pfaffian expression has an integrating factor, it has an infinite number of
integrating factors.

3) dL is neither exact nor does it have an integrating factor. Most Pfaffian
expressions are of this kind and cannot be used to construct state
functions: The differential dq is an exception.

4.7   PFAFFIAN EXPRESSIONS IN TWO VARIABLES

These differentials always have an integrating factor. To see this, consider the differential $dL = Xdx + Ydy$. It is assumed that dL is not exact, but it can be made exact by the integrating factor $\Lambda(x, y)$. Thus, the reciprocity relation $\partial(\Lambda X)/\partial y = \partial(\Lambda Y)/\partial x$ must hold, requiring that

$$\Lambda(\partial X/\partial y) + X(\partial \Lambda/\partial y) = \Lambda(\partial Y/\partial x) + Y(\partial \Lambda/\partial x) \qquad (4\text{-}29)$$

or

$$\Lambda[(\partial X/\partial y) - (\partial Y/\partial x)] = Y(\partial \Lambda/\partial x) - X(\partial \Lambda/\partial y) \qquad (4\text{-}30)$$

One can always find an integrating factor $\Lambda(x, y)$ that satisfies *Equation 4-29*. For example, for an ideal gas,

$$dq = \overline{C}_V(T)dT + (RT/\overline{V})d\overline{V}$$

Using *Equation 4-29* requires that

$$\Lambda(\partial \overline{C}_V/\partial \overline{V})_T + \overline{C}_V(\partial \Lambda/\partial \overline{V})_T = \Lambda R/\overline{V} + [RT/\overline{V}](\partial \Lambda/\partial T)_V \qquad (4\text{-}31)$$

where Λ is a function of T and V.

The simplest solution to this equation is to take $\Lambda(T,V) = 1/T$, yielding

$$1/T \times 0 + \overline{C}_V \times 0 = R/(T\overline{V}) + [RT/\overline{V}](-1/T^2) = 0 \qquad (4\text{-}32)$$

Thus, $1/T$ is an integrating factor, and so is any function $f(1/T)$ of $1/T$.

This example shows that $(1/T)dq$ is an exact differential and thus may be used as a basis for defining dS. However, the reasoning is faulty because this is a special case, namely, an application of a two-dimensional Pfaffian form, which always has an integrating factor. To serve as a basis for entropy, one has to show that dq has an integrating factor in any dimension. It happens to be true for dq, although not generally.

4.8   PFAFFIAN EXPRESSIONS IN MORE THAN TWO DIMENSIONS

In general, integrating factors in more than two variables do not exist. For example, even if we limit ourselves to three variables, $dL = Xdx + Ydy + Zdz$, the treatment would require solving three equations, such as Eq. 4-30. It would be very rare, indeed, to find a $\Lambda(x, y, z)$ for which the three eqations can be solved simultaneously.

In summary, Pfaffian differential expressions in two variables have integrating factors, but, when there are three or more variables, the Pfaffian forms have no integrating factors in *general*. However, under certain circumstances, some differential forms can admit of integrating factors, regardless of *dimensions*. How can these differential expressions be identified? The answer is provided by a Theorem proved by Carathéodory.

4.9 CARATHÉODORY'S THEOREM

If a Pfaffian expression $dL = \Sigma_i X_i \, dx_i$ *has the property for which, in the neighborhood of any point P, there are points that cannot be connected to P along curves that satisfy the equation* $dL = 0$, *then the Pfaffian expression has an integrating factor (in fact, an infinite number of infinite factors).* (We accept this without proof.)

Note: If dL stands for dq and there are points that cannot be reached from P along the locus dq = 0 (i.e., adiabatic path), then dq has an integrating factor. Carathéodory assumed this to be the case and postulated this as what is now known as Carathéodory's Principle (see Section 4.5).

4.10 ENTROPY—AXIOMATIC APPROACH

We now return to a discussion of the Second Law and the definition of entropy. First, let us consider *reversible* processes. In a reversible process, all external parameters (such as the generalized forces, X_i) that are part of dw, and so forth, are in effect equal to the internal forces of the system, defined by the equation of state. From the First Law we obtain

$$dq = dE - dw = (\partial E/\partial \theta)_x \, d\theta + \Sigma_i (\partial E/\partial x_i)_\theta \, dx_i - \Sigma_i X_i \, dx_i \qquad (4\text{-}33)$$

where the last term is an expression of the generalized work and θ is the empirical temperature. The first derivative of E is obviously the heat capacity (C). *Equation 4-33* is of Pfaffian form.

Carathéodory's Principle tells us that, in the vicinity of any state represented by a point P, there are states (points) that are inaccessible from P along the curve dq = 0. Thus, dq has an integrating factor Λ, which, in accordance with Carathéodory's Theorem, turns the inexact differential dq into the exact differential, $d\sigma$

$$\Lambda(\theta, x_1, \ldots, x_r) \, dq = d\sigma \qquad (4\text{-}34)$$

In the Axiomatic Approach to the Second Law, as we shall see, Λ is identified with $1/T$, where T is the absolute temperature, and the resulting exact differential, $d\sigma$, is identified with dS, where S is the entropy.

Before we continue with this discussion, two characteristics of Pfaffian forms need to be stated (although given without proof). If dL is an exact differential, or is a differential, $d\sigma$, made exact by an integrating factor, then L or σ are equal to a constant, defining a family of nonintersecting surfaces in the multipledimensional space. One can also prove that if a displacement from a point P on the surface to a neighboring point R satisfies the Pfaffian equation $(dL = 0)$ or $(d\sigma = 0)$, the point R will be on the same surface. Thus, if P is a point on the locus $dq = 0$ of a *reversible adiabat*, then a displacement to a point R along a Pfaffian equation will lie on the same adiabat.

We now proceed to show that, although Λ was introduced as a function of all the variables, $\theta, x_1, x_2, \ldots, x_r$, it is, in fact, a function of the empirical temperature θ only when the Pfaffian form is the differential dq. To demonstrate this, let us divide the system into two parts. The parts have different mechanical variables $x_1', x_2', \ldots$ and $x_1'', x_2'', \ldots$, and so forth but the same empirical temperature θ. The parts are in thermal equilibrium. By applying *Equation 4-34* separately to the parts, we have

$$\Lambda_1(\theta, x_1', x_2', \ldots, x_r')dq_1 = d\sigma_1 \tag{4-35}$$

and

$$\Lambda_2(\theta, x_1'', x_2'', \ldots, x_r'')dq_2 = d\sigma_2 \tag{4-36}$$

For the combined system

$$\Lambda(\theta, x_1', x_2', \ldots, x_1'', x_2'', \ldots)dq = d\sigma \tag{4-37}$$

Because $dq = dq_1 + dq_2$, we have

$$\Lambda^{-1}d\sigma = \Lambda_1^{-1}d\sigma_1 + \Lambda_2^{-1}d\sigma_2 \tag{4-38}$$

which holds for any arbitrary subdivision. This can only happen if Λ_1 and Λ_2 are independent of all the x variables but can depend only on the common empirical temperature, θ. Because $\Lambda = 1/T$, T can also depend only on the empirical temperature θ. Because T is a universal function of θ, we need to determine T for one system (say, an ideal gas) and we will know it for all systems in thermal equilibrium with it.

Rewriting *Equation 4-34* in terms of S and T and recalling the restriction to quasi-static processes in Carathéodory's Principle, we get the familiar expressions

$$dS = dq_{rev}/T \qquad (4\text{-}39a)$$

or

$$\Delta S = \int_{A}^{B} dq_{rev}/T \qquad (4\text{-}39b)$$

Let us look at the second part of the Second Law, namely, that for an isolated system: ΔS cannot decrease. Consider a system, whose states are defined by the variables T, S and other parameters, x. Assume, that reversible paths exist between two states (as drawn in Figure 4.7), depicting a reversible isotherm and a reversible adiabat. The latter must be perpendicular to the isotherm because $dq = TdS = 0$. We know from previous discussions that if we make a transition reversibly from point 1 along $dq = 0$, we wind up at a point along the adiabat $1 - 2$. Only irreversible transformations could possibly produce larger or smaller S values.

Actually, smaller S values are not possible. Without invoking the Clausius Principle, let us assume that we have a system that is adiabatically enclosed and that heat can flow spontaneously from one part of the system to the other. If T_1 is larger than T_2, we know from experience that the flow is from T_1 to T_2. We assume further that the parts are so large that the temperatures do not change perceptibly during the heat exchange (essentially, that

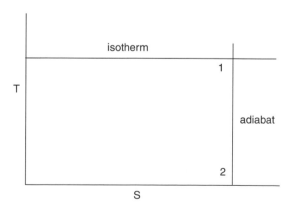

Figure 4.7 Illustration showing that points to the left of the (reversible) adiabat cannot be reached from a point on the adiabat. Points along the adiabat and to the right of it can be reached, showing that, for an isolated system, $\Delta S = 0$ for a reversible change and $\Delta S > 0$ for an irreversible change.

the parts are reservoirs). If $-q$ is the heat lost at T_1, obviously $+q$ is the heat gained at T_2. The change in entropy, however, is

$$\Delta S = -q/T_1 + q/T_2 = q(T_1 - T_2)/T_1 T_2 > 0 \qquad (4\text{-}40)$$

Thus, transitions can only be made along the adiabat or to the right of the adiabat. Points to the left are inaccessible.

4.11 ENTROPY CHANGES FOR NONISOLATED SYSTEMS

Let us now look at entropy changes for nonisolated systems. Let α be a system, which is not isolated, and β the surroundings. Together, they constitute an isolated system. We assume that α undergoes an irreversible change and that β is so large that changes are virtually imperceptible and can be regarded as reversible. Because the overall process is irreversible, we must have $\Delta S^\alpha + \Delta S^\beta > 0$, or

$$\Delta S^\alpha > -\Delta S^\beta \qquad (4\text{-}41a)$$

$$-\Delta S^\beta = -\int_A^B dq^\beta/T^\beta \qquad (4\text{-}41b)$$

Because the overall system is isolated, $dq^\alpha = -dq^\beta$ and

$$\Delta S^\alpha > \int_A^B dq^\alpha/T^\beta \qquad (4\text{-}41c)$$

We may combine *Equations 4-39b* and *4-41c* into a single formula

$$\Delta S \geq \int_A^B dq/T^\beta \qquad (4\text{-}42a)$$

where the $>$ sign refers to an irreversible transition and the $=$ sign to a reversible transition. The element of heat dq is the heat absorbed by the system, but the temperature T^β is the temperature of the surroundings. In a reversible change, there is no difference between the temperature of the system and of the surroundings, however, in an irreversible process, the temperatures are different, and, in fact, the temperature of the system is not well defined.

Equation 4-42a is most useful when applied to an isolated system or to the entire universe, treated as an isolated system, because $dq = 0$ and

$$\Delta S_{\text{isolated system}} \geq 0 \qquad (4\text{-}42b)$$

which is the same as *Equation 4-23*.

Note: Statements like *Equation 4-42b* led Clausius to assert that *"The energy of the universe remains constant, the entropy of the universe tends to a maximum."* This statement, sometimes referred to as the theory of the *thermal death* of the universe, was vehemently criticized by some scientists in the former Soviet Union. Some Scientists objected to extending thermodynamics, which is based on observations in a finite world, to an infinite world—in other words, to extending the subject beyond its proper domain. The theory of *thermal death* was attacked also on ideological ground. Bazarov (1964) asserted that the theory of the thermal death would "lead to religious superstition,—to the belief in God."

In chemical thermodynamics, the entropy of the system and its surroundings are often combined, simply because a numerical value can be assigned to the total and can readily be used to ascertain whether a process is reversible or irreversible. If the change is reversible, the total entropy is zero; if it is irreversible, the total entropy is greater than zero.

As an example, the transformation of supercooled water to ice, to be discussed shortly, is an irreversible process, but the entropy change is not positive, as might have been surmised. By adding the entropy of the surroundings, the total change becomes positive.

4.12 SUMMARY

We presented two approaches to the Second Law of Thermodynamics: traditional and axiomatic. Both approaches made use of statements that can be proved (theorems!) and statements that must be accepted without proof (principles!). The main objective of all of this is to develop the concept of entropy logically and rigorously. Not all steps in this development can be proved—some have to be postulated—thus the question arises: Why not postulate from the beginning all that is known about entropy, its existence, the properties, and so forth and circumvent even the mention of Carnot cycles or inaccessible adiabatic states? A number of such postulational approaches have been advanced. Here is one:

There exists a function, S, called the entropy, which is a state function. Its differential is related to the reversible heat, $dS = dq_{rev}/T$. For an isolated system, $\Delta S \geq 0$, where the = sign refers to a reversible transition and the > sign to an irreversible transition. This approach is no less fundamental or less logical than the traditional or axiomatic approaches. However, it postulates the existence and behavior of entropy not based on intuition or everyday experience. Contrast this to the other approaches. The Clausius Principle, for example, without its trimmings, asserts that heat cannot of itself flow from a colder to a hotter body, which is an experience common to all. The inaccessibility of some adiabatic states demanded by Carathéodory's Principle is not commonly experienced, but it is a statement consistent with the Clausius Principle.

4.13   SOME APPLICATIONS OF THE SECOND LAW

4.13.1   Reversible Processes (PV Work Only)

$$dS = dq_{rev}/T$$

$$= dE/T + (P/T)dV$$

$$= 1/T\{(\partial E/\partial T)_V \, dT + [(\partial E/\partial V)_T + P]dV\} \tag{4-43}$$

$$= (C_V/T)dT + 1/T[(\partial E/\partial V)_T + P]dV \tag{4-44}$$

Thus

$$(\partial S/\partial T)_V = C_V/T \tag{4-45}$$

$$(\partial S/\partial V)_T = 1/T[(\partial E/\partial V)_T + P] = (\partial P/\partial T)_V \tag{4-46}$$

upon applying the Thermodynamic Equation of State, $(\partial E/\partial V)_T = T(\partial P/\partial T)_V - P$

Also,

$$(\partial^2 P/\partial T^2)_V = [\partial/\partial T(\partial S/\partial V)_T]_V = \partial^2 S/\partial T \partial V = \partial^2 S/\partial V \partial T$$

$$= [\partial/\partial V(C_v/T)]_T = 1/T[(\partial C_V)/\partial V)_T] \tag{4-47}$$

Similar relations exist for $(\partial S/\partial T)_P$, $(\partial S/\partial P)_T$, and so forth (see Problem Set III in Appendix I).

$$\Delta S = \int (C_V/T)dT + 1/T \int [(\partial E/\partial V)_T + P]dV \tag{4-48}$$

For one mole of an ideal gas, with constant $\overline{C}_V$,

$$\Delta \overline{S} = \overline{C}_V \ln(T_2/T_1) + R \ln(\overline{V}_2/\overline{V}_1) \tag{4-49}$$

Equivalently,

$$\Delta \overline{S} = \overline{C}_V \ln(T_2/T_1) + R \ln(T_2 P_1/P_2 T_1)$$

$$= (\overline{C}_V + R) \ln(T_2/T_1) + R \ln(P_1/P_2)$$

$$= \overline{C}_P \ln(T_2/T_1) + R \ln(P_1/P_2) \tag{4-50}$$

(This can be obtained directly.)

$$dS = dq_{rev}/T = dH/T - (V/T)dP \qquad (4\text{-}51)$$

$$\Delta S = \int C_P dT/T + 1/T \int [(\partial H/\partial P)_T - V]dP$$

$$= C_P \ln(T_2/T_1) + 1/T \int [V - T(\partial V/\partial T)_P - V]dP \qquad (4\text{-}52)$$

For one mole of an ideal gas,

$$\Delta \overline{S} = \overline{C}_P \ln(T_2/T_1) - R \ln(P_2/P_1)$$

For vaporization (constant T, P)

$$\Delta S = \Delta H_{vap}/T \left(= \int dq_{rev}/T \right) \qquad (4\text{-}53)$$

Note: We have made frequent use of the so-called Thermodynamic Equations of State. The easiest way to derive them is from the Maxwell relations, to be discussed in Chapter 5. However, we can also obtain them from the general expressions of dS (*Eqs.* 4-43 and 4-51) by observing that dS is an exact differential and therefore the reciprocity relations hold. Applying these we readily obtain the Thermodynamic Equations of State.

4.13.2 Irreversible Processes

If the transition is irreversible, then one must devise a reversible path.

Example: Calculate ΔS for the transformation of 1 mol of supercooled water at $-10°C$ to ice at $-10°C$. This is a nonequilibrium irreversible path, which can be evaluated by choosing the following reversible path.

$$H_2O(l, 0°C) \rightarrow_{rev} H_2O(s, 0°C)$$
$$\uparrow_{rev} \qquad\qquad\qquad \downarrow_{rev} \qquad\qquad (4\text{-}54)$$
$$H_2O(l, -10°C) \rightarrow_{irrev} H_2O(s, -10°C)$$

Note: It is obvious by now that entropy is basic to any discussion of natural changes in thermodynamics, but to use it, one must consider both the system and the surroundings except when the system is isolated. But there ways to take account of the effect of the surrounding without using it. This is accomplished by the use of two new thermodynamic finctions, the Helmholtz free energy and the Gibbs free energy, to be discussed in the next chapter.

CHAPTER 5

USEFUL FUNCTIONS: THE FREE ENERGY FUNCTIONS

Combining the First and Second Laws, namely, $dE = dq - PdV$ and $dS = dq_{rev}/T$, gives

$$dE = TdS - PdV \qquad (5\text{-}1)$$

Note: It is implied in this and the subsequent sections that P and T are well defined, i.e., pertaining to the properties of the system.

This expression of dE in terms of dS and dV is extremely simple and S and V are said to be the "natural variables" of E. However, they are not the most convenient variables in treating chemical problems. The most convenient variables are T and P. It is, of course, possible to write dE as a function of T and V, namely $dE = C_v dT + [T\ (\partial P/\partial V)_T - P]\ dV$, which we used before; however the coefficients of dV and dT are not as simple as those of dV and dS.

The question obviously arises: "Are there thermodynamic state functions whose natural variables are P and T or V and T or S and P, etc., and how can we find them?"

In elementary discussions, it is common practice to define new functions by combining E, S, and V and to show that their differentials have the desired characteristics. The new functions are as follows:

Thermodynamics and Introductory Statistical Mechanics, by Bruno Linder
ISBN 0-471-47459-2 © 2004 John Wiley & Sons, Inc.

- *Enthalpy*: $H = E + PV$ (variables S, P)

$$dH = TdS + VdP$$

- *Helmholtz Free Energy*: $A = E - TS$ (variables T, V)

$$dA = -SdT - PdV$$

- *Gibbs Free Energy*: $G = H - TS$ (variables T, P)

$$dG = -SdT + VdP$$

The generation of the above combinations appears to be arbitrary and do not exhaust all possibilities. There are other functions that are also simple and useful. A general mathematical technique exists for generating all such functions. This leads to the topic of Legendre Transformations.

5.1 MATHEMATICAL INTERLUDE IV: LEGENDRE TRANSFORMATIONS

Equation 5-2 is the well-known representation of a straight line in the x, y frame. ξ and ϕ are, respectively, the slope and the intercept of the line.

$$y = \xi x + \phi \tag{5-2}$$

If the line is curved, the equation can be represented as

$$y = \xi(x)x + \phi(\xi) \tag{5-3}$$

Here, $\xi = dy/dx$. The curve (*Eq. 5-3*) can be completely described by specifying x and the corresponding y or by specifying the slope ξ and the corresponding intercept ϕ; that is (as is obvious)

$$\phi(\xi) = y - \xi x$$

or

$$\phi(\xi) = y(x) - x(dy/dx) \tag{5-4}$$

The variable ξ is considered to be the independent variable, and the function $\phi(\xi)$ is the Legendre transformation of y.

Generalization to more than one variable gives

$$\phi(\xi_1 \cdots \xi_r) = Y(x_i, \ldots, x_r) - \Sigma_i x_i \partial y / \partial x_i = Y - \Sigma_i x_i \xi_i \qquad (5\text{-}5)$$

The Legendre transformation transforms a function Y of the variables x_i in terms of the slopes of Y with respect to x_i.

5.1.1 Application of the Legendre Transformation

Given the function $E = E(S,V)$, find a function whose natural variables are T and P. Note that T and P are the slopes of E with respect to S and V; thus the Legendre transformation should be applicable.

$$\phi(T,P) = E(S,V) - S(\partial E/\partial S)_V - V(\partial E/\partial V)_S \qquad (5\text{-}6)$$
$$\phi = E - ST - V(-P) = E + PV - TS \qquad (5\text{-}7)$$

This is, of course, the Gibbs free energy G.

It is always a good idea to verify that ϕ is indeed a function of T and P. Looking at the differential $d\phi$ and using *Equation 5-7* yields

$$d\phi = dE - TdS - SdT + PdV + VdP \qquad (5\text{-}8)$$

Substituting

$$dE = TdS - PdV \qquad (5\text{-}9)$$

gives

$$d\phi = -SdT + VdP \qquad (5\text{-}10)$$

which proves that ϕ is indeed a function of the variables T and P.

Given the function $E(S,V)$, find a function, ϕ, whose natural variables are T and V. This is an example of a one-variable problem because V is held fixed throughout. Thus, ignoring differentiation with respect to V gives

$$\phi(T,V) = E(S,V) - S(\partial E/\partial S)_T = E - TS \qquad (5\text{-}11)$$

(Obviously, the function is the Helmholtz Free Energy, A.)

To verify that ϕ is a function of T and V, we write

$$d\phi = dE - TdS - SdT$$
$$= TdS - PdV - TdS - SdT$$
$$= -SdT - PdV \qquad (5\text{-}12)$$

5.2 MAXWELL RELATIONS

From the fundamental relations on the left of the equations below, we obtain at once the Maxwell relations on the right, simply by making use of the reciprocity relations (see *Eq. 3-12* in Chapter 3).

1. $dG = -SdT + VdP$	$-(\partial S/\partial P)_T = (\partial V/\partial T)_P$	(*5-13a*)
2. $dA = -SdT - PdV$	$(\partial S/\partial V)_T = (\partial P/\partial T)_V$	(*5-13b*)
3. $dH = TdS + VdP$	$(\partial T/\partial P)_S = (\partial V/\partial S)_P$	(*5-13c*)
4. $dE = TdS - PdV$	$-(\partial P/\partial S)_V = (\partial T \partial P)_V$	(*5-13d*)

The last two Maxwell Relations are less important than the first two.

Note: The easiest way to derive the Thermodynamic Equations of State (see Section 3.7) is to obtain them via the Maxwell Relations. Combining $(\partial E/\partial V)_T = T(\partial S/\partial V)_T$ with the right-hand side of *Equation 5-13b* and $(\partial H/\partial P)_T = T(S/\partial P)_T$ with the right-hand side of *Equation 5-13a* yields the Thermodynamic Equations of State (*Eqs. 3-43* and *3-44*).

5.3 THE GIBBS-HELMHOLTZ EQUATIONS

From the relations on the left-hand of *Eq. 5-13a* and *Eq. 5-13b*, one immediately obtains a set of equations, known as the Gibbs-Helmholtz equations. These equations relate the temperature derivatives of A and G to E and H, respectively

$$[\partial/\partial T(G/T)]_P = 1/T(-S) - [H - TS]/T^2$$
$$= -H/T^2 \qquad (5\text{-}14a)$$

or

$$[\partial(G/T)/\partial(1/T)]_P = -T^2[\partial(G/T)/\partial T]_P = H \qquad (5\text{-}14b)$$
$$[\partial/\partial T(A/T)]_V = 1/T(\partial A/\partial T)_V - A/T^2$$
$$= 1/T(-S) - [E - TS]/T^2$$
$$= -E/T^2 \qquad (5\text{-}15a)$$

or

$$[\partial(A/T)/\partial(1/T)]_V = -T^2[\partial(A/T)/\partial T]_V = E \qquad (5\text{-}15b)$$

5.4 RELATION OF ΔA AND ΔG TO WORK: CRITERIA FOR SPONTANEITY

5.4.1 Expansion and Other Types of Work

We have seen before that useful relations exist between ΔE, ΔH, and q, namely, $\Delta E = q_V$ and $\Delta H = q_p$, provided that there is no work other than pressure-volume (PV) work. Other relations, even more useful, exist between ΔA, ΔG, and w. These relations entail not only pressure-volume work but also other types of work, which we designate as "other" work: $w_{tot} = w_{PV} + w_{other}$. Some authors refer to this work as "net" work or "useful" work. It is assumed that under all circumstances P and T are well defined, that is, that they are properties of the system. From the Clausius Inequality $\oint dq/T \leq 0$ (*Eq. 4-19*) and the relation $_A\int^B dq/T \leq S_B - S_A$ (*Eqs. 4-20b and 4-22*), we deduce the following relations:

1) *Constant T*:

$$1/T \int dq = (1/T) q_{A \to B} \leq S_B - S_A \qquad (5\text{-}16)$$

or

$$q_{A \to B} \leq TS_B - TS_A \qquad (5\text{-}17)$$

Adding $w_{A \to B}$ to both sides of the equation yields

$$q_{A \to B} + w_{A \to B} \leq TS_B - TS_A + w_{A \to B} \qquad (5\text{-}18)$$

Thus,

$$E_B - E_A \leq TS_B - TS_A + w_{A \to B} \qquad (5\text{-}19a)$$
$$(E_B - TS_B) - (E_A - TS_A) \leq w_{A \to B} \qquad (5\text{-}19b)$$

or

$$\Delta A_T = A_B - A_A \leq w_{tot} \qquad (5\text{-}20)$$

2) *Constant T and V*:

$$w_{tot} = w_{other}$$

and

$$\Delta A_{T,V} \leq w_{other} \qquad (5\text{-}21a)$$

If there is no "other" work and V is constant,

$$\Delta A_{T,V} \leq 0 \qquad (5\text{-}21b)$$

3) *Constant T and P*:

$$w_{tot} = -P(V_B - V_A) + w_{other}$$

Using *Equation 5-20a* gives

$$(A_B + PV_B) - (A_A + PV_A) \leq w_{other} \qquad (5\text{-}22)$$

or

$$\Delta G_{T,P} = G_B - G_A \leq w_{other} \qquad (5\text{-}23a)$$

In the absence of "other" work, the condition is

$$G_{T,P} \leq 0 \qquad (5\text{-}23b)$$

Relation (5-23a) is useful because it provides a means for determining ΔG from work measurements. If w_{other} is electrical work, generated by a galvanic cell, for example, it is related to the cell potential in the manner $w_{other} = -v\mathcal{F}\mathcal{E}^0$, where v is the number of moles of electrons, $\mathcal{F}$ is the Faradys's constant (96,486 k Coulomb/mol), and $\mathcal{E}^0$ is the cell potential. Thus, by measuring the cell potential reversibly, one readily obtains $\Delta G_{T,P}$.

If only PV-work is present, *Equation 5-23a* reduces to (*5-23b.*)

It is now clear why the Gibbs and the Helmholtz energies are called "free" energies. They are part of the energy of the system, which is free to do work.

5.4.2 Comments

The free energy expressions given by *Equations 5-19b* and *5-23b* are particularly revealing because they state that, if the change is negative, the transition is irreversible. As noted before (and to be elaborated on more fully in Chapter 7), spontaneous processes are irreversible. It is noted that both free energy expressions are restricted, ΔA to constant T and V and ΔG to constant T and P. This is in contrast to the change in entropy (*Eq. 4-42b*), which is restricted to an isolated system but is frequently replaced by the whole universe, implying that the *system* and the entire *surroundings* constitute an *isolated* system. The condition for reversibility in terms of entropy changes is often written as $\Delta S_{universe} \geq 0$. The question now arises: Why is it necessary to include the surroundings when using the entropy criterion but not the Helmholtz or Gibbs free energies? The change in the entropy, $\Delta S_{universe}$, refers to the *system* and the *environment*. The change in the Gibbs and Helmholtz free energies refer only to the *system*. The following analysis may be instructive.

It was shown previously (in Chapter 3) that, when a system is capable of doing only PV work, $\Delta H = q_P$. If there is also "other" work, obviously $\Delta H = q_P + w_{other}$. Recall that, in deriving the Clausius Inequality (*Eq. 4-19*), the system was taken from state A to state B by an irreversible path and returned from B to A by a reversible path. It was tacitly assumed that this is always possible; in fact, the inequality $\Delta S_{isolated} \geq 0$ is based on that assumption. If the isolated system consists of the entire universe, it is reasonable to assume that changes in the environment can proceed quasi-statically, since the environment is so enormous. When the system exchange heat with the surroundings, we must have $q_{sys} = -q_{surr}$. Therefore, $q_{surr} = -\Delta H_{sys} + w_{other}$. The entropy change of the surroundings at constant T and P is $\Delta S_{surr} = -\Delta H_{sys}/T + w_{other}/T$ and the total entropy change of system and surroundings is $\Delta S_{tot} = \Delta S_{sys} + \Delta S_{surr} = \Delta S_{sys} - \Delta H_{sys}/T + w_{other}/T = -\Delta G_{sys}/T + w_{other}/T$. Thus, although ΔG is clearly defined in terms of ΔS and ΔH—of the *system*, it is obvious that it is also related to the *total* entropy change, namely, the change in the *system* and the *surroundings*. In summary, when the entropy is used to establish spontaneous or natural transformations, the total entropy change of system and surroundings must be used. At constant T and P, the entropy change of the *surroundings* can be replaced by $-\Delta H_{sys}/T + w_{other}/T$, provided that the heat exchange of the surroundings is treated as reversible and the entire transformation can be written in terms of properties of the *system*. Notice that because $\Delta S_{tot} = \Delta S_{universe} \geq 0$ we obtain $\Delta G_{T,P} \leq w_{other}$.

5.5   GENERALIZATION TO OPEN SYSTEMS AND SYSTEMS OF VARIABLE COMPOSITION

5.5.1   Single Component System

Let the superscript bar denote a molar value at constant P and T. It is known from experience that $C_P(T,P; n) = n\overline{C}_P(T,P)$. Also, $G(T,P; n) = n\overline{G}(T,P)$, $S(T,P; n) = n\overline{S}(T,P)$, etc. Accordingly,

$$dG = nd\overline{G} + \overline{G}dn \qquad (5\text{-}24a)$$

$$= n(-\overline{S}dT + \overline{V}dP) + \overline{G}dn \qquad (5\text{-}24b)$$

$$= -SdT + VdP + \overline{G}dn \qquad (5\text{-}24c)$$

or

$$\overline{G} = (\partial G/\partial n)_{T,P} = G/n \qquad (5\text{-}25)$$

5.5.2 Multicomponent Systems

$$G = G(T, P; n_1, \ldots, n_r)$$

$$dG = (\partial G/\partial T)_{P;\,\underline{n}_i} dT + (\partial G/\partial P)_{T;\underline{n}_i} dP + \Sigma_i (\partial G/\partial n_i)_{T,P;\,n_{j\neq i}} dn_i \quad (5\text{-}26a)$$

$$dG = -SdT + VdP + \Sigma_i (\partial G/\partial n_i)_{T,P;n_{j\neq i}} dn_i \qquad (5\text{-}26b)$$

The bar under the n_i serves to denote that all n_i variables are held constant. The symbol $n_{j\neq i}$ denotes that all mole numbers except n_i are to be held constant.

Similarly,

$$E = E(S, V; n_1, \ldots, n_r)$$

$$dE = (\partial E/\partial S)_{V;\,\underline{n}_i} dS + (\partial E/\partial V)_{S;\,\underline{n}_i} dV + \Sigma_i (\partial E/\partial n_i)_{S,V;\,n_{j\neq i}} dn_i \quad (5\text{-}27a)$$

$$= TdS - PdV + \Sigma_i (\partial E/\partial n_i)_{S,V;\,n_{j\neq i}} dn_i \qquad (5\text{-}27b)$$

But also $G = E - TS + PV$ and so

$$dG = dE - TdS - SdT + PdV + VdP \qquad (5\text{-}28a)$$

Substituting *Eq. 5-27b* for dE gives

$$dG = -SdT + VdP + \Sigma_i (\partial E/\partial n_i)_{S,V;\,n_{j\neq i}} dn_i \qquad (5\text{-}28b)$$

Comparing *Eqs. 5-26b* and *5-28b* shows that

$$(\partial G/\partial n_i)_{T,P;\,n_{j\neq i}} = (\partial E/\partial n_i)_{S,V;\,n_{j\neq i}} \qquad (5\text{-}28c)$$

In similar fashion, we can show that

$$(\partial G/\partial n_i)_{T,P;\,n_{j\neq i}} = (\partial H/\partial n_i)_{S,P;\,n_{j\neq i}} = (\partial A/\partial n_i)_{T,V;\,n_{j\neq i}} \qquad (5\text{-}28d)$$

5.6 THE CHEMICAL POTENTIAL

The partial derivatives in the preceding section transcend the ordinary properties of partials and deserve a special name. They are called *chemical potentials* and are denoted by the symbol μ_i. Thus, although the chemical potential is generally referred to as the partial molar Gibbs free energy, it is really also the partial molar Helmholtz free energy, the partial molar enthalpy, and the partial molar energy. Using the common symbol μ_i, the differentials of the thermodynamic functions can now be expressed as

$$dE = TdS - PdV + \Sigma_i \mu_i \, dn_i \qquad\qquad (5\text{-}29a)$$

$$dH = TdS + VdP + \Sigma_i \mu_i \, dn_i \qquad\qquad (5\text{-}29b)$$

$$dA = -SdT - PdV + \Sigma_i \mu_i \, dn_i \qquad\qquad (5\text{-}29c)$$

$$dG = -SdT + VdP + \Sigma_i \mu_i \, dn_i \qquad\qquad (5\text{-}29d)$$

5.7 MATHEMATICAL INTERLUDE V: EULER'S THEOREM

The standard definitions of Intensive and Extensive Variables are:

- *Intensive Variables* are independent of the mass of material.
- *Extensive Variables* are dependent on the mass of the material.

A mathematical definition (following Wall, 1965) states:

An extensive variable is a homogeneous function of first degree in the masses of the material, with such homogeneity being fulfilled while all intensive variables are held constant. (An intensive property is a homogeneous function of degree zero in the masses.)

- *Homogeneous function of degree n*: $f(x, y, z)$ is homogeneous of degree n in the variables x, y, z if

$$f(\lambda x, \lambda y, \lambda z) = \lambda^n f(x, y, z) \qquad\qquad (5\text{-}30)$$

- *Euler's Theorem*: if $f(x, y, z)$ is homogeneous of degree n, then

$$x(\partial f / \partial x) + y(\partial f / \partial y) + z(\partial f / \partial z) = nf(x, y, z) \qquad\qquad (5\text{-}31)$$

- *Proof*: differentiate *Eq. 5-30* with respect to λ. This gives

$$n\lambda^{n-1} f(x, y, z) = [\partial f / \partial(\lambda x)][\partial(\lambda x)/\partial\lambda]$$
$$+ [\partial f / \partial(\lambda y)][\partial(\lambda y)/\partial\lambda] + [\partial f / \partial(\lambda z)][\partial(\lambda z)/\partial\lambda] \qquad (5\text{-}32)$$
$$= x\partial f / \partial(\lambda x) + y\partial f / \partial(\lambda y) + z\partial f / \partial(\lambda z) \qquad\qquad (5\text{-}33)$$

Setting $\lambda = 1$ obtains

$$nf(x, y, z) = x\partial f / \partial x + y\partial f / \partial y + z\partial f / z \qquad\qquad (5\text{-}34)$$

If $n = 1$, f is homogeneous of degree 1; if $n = 0$, f is homogeneous of degree 0.

EXAMPLES

1. $V = V(T, P; n_1, \ldots, n_r)$ is homogeneous of degree 1 in the extensive variables n_1, n_r. Thus, $V = V(T, P; \lambda n_1, \ldots, \lambda n_r)$. The intensive variables T, P must be held constant. Thus, by Euler's Theorem

$$V = \Sigma_i n_i (\partial V / \partial n_i)_{T,P;n_{j \neq i}} \qquad (5\text{-}35)$$

2. $A = A(T, V; n_1, \ldots, n_r)$ is homogeneous of degree 1 in the extensive variables V, $n_1, \ldots, n_r$. Thus, $A = A(T, \lambda V; \lambda n_1, \ldots, \lambda n_r)$. Holding T constant gives

$$A = V(\partial A / \partial V)_{T;n_i} + \Sigma_i (\partial A / \partial n_i)_{V;n_j \neq i} \qquad (5\text{-}36)$$

5.8 THERMODYNAMIC POTENTIALS

In *Equations 5-29a, b, c,* and *d,* we obtained expressions for the differentials dE, dH, dA, and dG in terms of all the system variables. We are now in a position to derive expressions for the quantities E, H, A, and G themselves rather than for their differentials. They are called *thermodynamic potentials* because each gives a complete thermodynamic description of the system, as will be seen shortly.

First, consider the function $dE = TdS - PdV + \Sigma_i \mu_i dn_i$ (*Eq. 5-29a*). To obtain E, one might be tempted to integrate each term indefinitely and write

$$E = TS - PV + \Sigma_i \mu_i n_i \qquad (5\text{-}37)$$

Normally, the result would be incorrect because no account is taken of the integration constants. The result happens to be correct, as can be verified by applying Euler's Theorem. Because E is homogeneous of first degree in the variables S, V, and all the n_i, Euler's Theorem predicts the validity of *Equation 5-37.*

Other thermodynamic potentials are not that simply related to their differential forms. Application of Euler's Theorem shows that

$$H = TS + \Sigma_i \mu_i n_i \qquad (5\text{-}38)$$

$$A = -PV + \Sigma_i \mu_i n_i \qquad (5\text{-}39)$$

$$G = \Sigma_i \mu_i n_i \qquad (5\text{-}40)$$

Note: If we know E, we can get all other functions by applying Legendre Transformations without using Euler's Theorem.

EXAMPLES

1. Given $E = E(S,V; n_1, \ldots, n_r)$, find a function whose natural variables are $T,V; n_1, \cdots, n_r$. The Legendre Transformation is

$$\phi(T, V; n_1, \ldots, n_r) = E - S(\partial E/\partial S)_{V;\underline{n}_i} = E - TS$$

$$= -PV + \Sigma_i\mu_i n_i \qquad (5\text{-}41)$$

upon substitution of *Eq. 5-37* for E.
   This can be checked as follows:

$$d\phi = dE - TdS - SdT \qquad (5\text{-}42a)$$

$$= TdS - PdV + \Sigma_i\mu_i n_i - TdS - SdT \qquad (5\text{-}42b)$$

$$= -SdT - PdV + \Sigma_i\mu_i dn_i \qquad (5\text{-}42c)$$

In other words, ϕ (obviously A) is indeed a function of T,V and the n_i.

2. From E $(S,V; n_i, \ldots, n_i)$, obtain a function $\phi = \phi(T,P; \mu_1 \cdots \mu_r)$. Notice that all the variables are intensive.

$$\phi = E - S(\partial E/\partial S)_{V;\underline{n}_i} - V(\partial E/\partial V)_{S;\underline{n}_i} - \Sigma_i n_i(\partial E/\partial n_i)_{S,V;\,n_{j\neq i}} \qquad (5\text{-}43a)$$

$$= E - TS + PV - \Sigma_i\mu_i n_i = 0 \qquad (5\text{-}43b)$$

The potential ϕ is peculiar in the sense that it is always equal to zero and so is its differential

$$d\phi = -SdT + VdP - \Sigma_i n_i d\mu_i = 0 \qquad (5\text{-}44)$$

from which several useful relations can be obtained.
   At constant μ_i

$$-SdT + VdP = 0 \qquad (5\text{-}45)$$

$$(\partial P/\partial T)_{\mu_i} = S/V \qquad (5\text{-}46)$$

At constant T and $\mu_{j\neq i}$

$$VdP - n_i d\mu_i = 0 \qquad (5\text{-}47)$$

$$(\partial P/\partial\mu_i)_{T;\mu_{j\neq i}} = \mu_i/V \qquad (5\text{-}48)$$

Because the functions H, A, and G can be obtained from E by a Legendre Transformation, which is a mathematical transformation, they must contain the same thermodynamic information as E.

The function E is expressed in terms of the extensive variables S, V, and all the n_i. The remaining variables are the derivatives of E with respect to these variables. This is precisely what the Legendre transformation accomplishes: it transforms a function in terms of its derivatives.

The independent variables, in terms of which the thermodynamic potentials are expressed, are referred to as *characteristic variables*. The best way to determine this is from the differential forms, such as *Eqs. 5-39, 5-42c, 5-44,* and others.

Note: Generalization of the thermodynamic potentials to multicomponent and open systems was accomplished by considering the combined First and Second Laws. Can the First and Second Law be generalized individually so as to apply to a multicomponent or an open system? The answer is yes, but the derivation will depend on an additional assumption, namely, that energy is additive.

Consider a system of energy E and volume V. Set up a second system of infinitesimal amount dn_i of substance i (see Figure 5.1). The two systems are in thermal equilibrium, mechanical equilibrium, and matter equilibrium, insofar as matter flow through membrane permeable to i is concerned. We incorporate the small system into the large system in two steps.

First, extend the boundary of the large system to incorporate the small system. Second, push in the piston (reversibly) until the original volume is restored. This procedure is first carried out adiabatically.

- *Step 1*: $dE = \bar{E}_i \, dn_i; \quad dS = \bar{S}_i dn_i$
- *Step 2*: $dE = P\bar{V}_i dn_i; \quad dS = 0$ (because it is an adiabatic change) $(5\text{-}49)$

Overall, $dE = (\bar{E}_i + P\bar{V}_i) \, dn_i; \quad dS = \bar{S}_i \, dn_i$

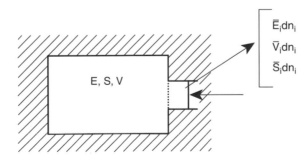

Figure 5.1 Schematic representation of small system to be incorporared into larger system.

This process is repeated with r different species which results in

$$dE = \Sigma_i(\overline{E}_i + P\overline{V}_i)\,dn_i; \quad dS = \Sigma_i\overline{S}_i dn_i \tag{5-50}$$

The adiabatic enclosure is now replaced by a diathermal boundary, which allows reversible exchange of heat and work (PV-work) between the system and the surrounding, yielding

- *First Law:* $dE = dq + dw + \Sigma_i(\overline{E}_i + P\overline{V}_i)dn_i$ $\tag{5-51}$

- *Second Law:* $dS = \dfrac{dq_{rev}}{T} + \Sigma_i\overline{S}_i dn_i$ $\tag{5-52}$

Note that since $\overline{G}_i = \overline{E}_i + P\overline{V}_i - T\overline{S}_i = \mu_i$, the combined First and Second Law expressions give the old formula $dE = TdS - PdV + \Sigma_i\mu_i dn_i$.

CHAPTER 6

THE THIRD LAW OF THERMODYNAMICS

Many people regard the Third Law as a curiosity of no great practicality. For chemists, the Third Law is a statement of great importance. For example, if one wishes to obtain ΔS or ΔG for the reaction

$$C(s) + 2H_2(g) \rightarrow CH_4(g) \qquad (6\text{-}1a)$$

at T = 298 K and P = 1 atm, it is easy enough to measure ΔH calorimetrically; however, to obtain ΔS or ΔG calorimetrically may be very difficult if not impossible. Recall that the determination of ΔS requires a reversible path, which is often hard to construct and may turn out that the reaction rate along that path is so slow as to be virtually immeasurable. With the help of the Third Law, especially in the form of so-called Nernst's Heat Theorem, the determination is relatively easy.

To be specific, suppose one wishes to determine ΔS for the reaction displayed in *Equation 6-1a*. One could imagine a reversible path that would take the reactants C and H_2 from 1 atm pressure to the equilibrium pressure, P^*, where they would react to form CH_4 and return CH_4 from P^* to 1 atm. But this may be impossible because P^* is very small and no reaction is likely to take place. Another possibility is to choose a path that takes the reactants

---

Thermodynamics and Introductory Statistical Mechanics, by Bruno Linder
ISBN 0-471-47459-2 © 2004 John Wiley & Sons, Inc.

reversibly from T to $T_0 = 0$ K, producing the product CH_4 and returning from T_0 to T, all at P = 1 atm (see *Eq. 6-1b*).

$$C(s, 1\,atm, T = 0) + H_2(s, 1\,atm, T = 0) \rightarrow CH_4(s, 1\,atm, T = 0)$$

$$\uparrow \qquad\qquad\qquad \uparrow \qquad\qquad\qquad \downarrow \qquad\qquad (6\text{-}1b)$$

$$C(s, 1\,atm, T) + \qquad H_2(g, 1\,atm, T) \qquad \rightarrow CH_4(g, 1\,atm, T)$$

6.1 STATEMENTS OF THE THIRD LAW

There are several statements of the Third Law, none completely satisfactory from a phenomenological point of view. They come under several headings:

1) *Nernst Heat Theorem*:

$$\lim_{T \to 0} \Delta S = 0 \quad \text{or} \quad \Delta S_0 = 0 \qquad\qquad (6\text{-}2)$$

This statement is based on observations by Richards (1902), who found that for several galvanic reactions formula (6-2) holds. Nernst thought that it should be universally true.

2) *Planck*:

$$\lim_{T \to 0} S = 0 \quad \text{or} \quad S_0 = 0 \qquad\qquad (6\text{-}3)$$

Planck assumed this relation to hold for every "homogeneous chemical body"

3) *G. N. Lewis*

Every substance has a finite positive entropy, but at the absolute zero of temperature, the entropy may become zero and does so in the case of a perfect crystalline solid.

4) *Unattainability of Absolute Zero*

This is another form of the Third Law, which some regard as most satisfactory, although it does not have the practicality of the previous statements. This form says that no system can be cooled down to absolute zero. It can be shown that this statement is consistent or leads to the Nernst Heat Theorem.

There are problems with each of the statements.

The Nerst Heat Theorem is neither a theorem nor has it anything to do with heat. Furthermore, it is too general.

Planck's statement is not based on experimental observations. One cannot measure entropy itself, only entropy changes. Furthermore, it is too general and does not apply to all homogeneous substances.

The Lewis statement has similar shortcomings, and is, moreover, limited to "perfect" crystalline solids. However, how does one know when a crystalline solid is "perfect"?

The entropy of a substance can be determined two ways: calorimetrically (to be discussed in Section 6.1) and statistically (i.e., by means of statistical mechanics). In most instances, the results are the same, but there are exceptions. In all cases where differences have been observed, the discrepancies have been traced to the presence of "frozen-in" structures. Molecules such as CO, NNO, etc. have dipole moments, which should all point in the same direction as the temperature goes to zero. With these molecules, however, the dipole moments are so small that they do not line up as the substance is cooled down to $T = 0$. The random structures are "frozen-in," and the solid is not "perfect." In other words, the systems are not in "*true thermodynamic equilibrium*," and the measured entropies close to zero will not coincide with statistical entropies, which are calculated on the basis of true thermodynamic equilibrium. From this standpoint, a perfect crystalline solid may be defined as one that is in true thermodynamic equilibrium. But it must be rememberd that, to ascertain whether a solid is in true thermodynamic equilibrium, one must go outside the realm of thermodynamics and seek information from other disciplines, such as statistical mechanics, structural measurements, and so forth.

Because of the exceptions, the Third Law has not been universally accepted as a Law on par with the First and Second Laws. Although the so-called exceptions can be accounted for, the explanations are based on extrathermodynamic arguments; for this reason, some people maintain that the Third Law should not be a part of the Laws of Thermodynamics.

The most common use of the Third Law is in the form of Planck's or Lewis's statements, which assume that $S_0 = 0$ (under any condition of pressure). The Third Law allows us to determine the actual entropy of pure substances, rather than entropy differences. If $\Delta S = S(T) - S(0)$ represents the entropy difference between the temperature T and $T = 0$ K and because $S(0)$ is zero by the Third Law, the actual $S(T)$ is the same as ΔS.

The determination of entropy at a finite temperature, T, is generally obtained from experimental heat capacity and transition enthalpy (solid-liquid, liquid-vapor, etc.) data (See Fig. 6.1). At very low temperatures, where the substances are almost always solids, there is virtually no

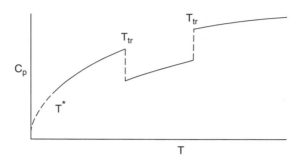

Figure 6.1 Schematic representation of the variation of C_P with T. The symbol T_{tr} represents solid-liquid and liquid-vapor transition temperatures; the low temperature dotted line is extrapolated.

difference between C_P and C_V, and the heat capacity at these low temperatures is generally denoted as C. The entropy, S, can be obtained from a plot of C_P/T vs. T, to which transition entropies, $\Delta H_{trans}/T$, are added. The trouble is that C_P cannot be measured at very low temperatures (say below 10 K). What is normally done is to estimate C from the Debye-Cube Law, $C = aT^3$ (which Debye derived statistically), where the parameter a is a constant. If T^* is the lowest T for which C can be measured, then

$$S_T = \int_0^{T^*} (C_{Deb}/T)dT + \sum \int_{T^*}^{T_{tran}} (C/T)dT + \sum \Delta H_{tran}/T_{tran} \quad (6\text{-}4a)$$

$$= 1/3aT^{*3} \qquad + \qquad '' \qquad + \qquad '' \qquad (6\text{-}4b)$$

6.2 ADDITIONAL COMMENTS AND CONCLUSIONS

Nernst thought that the Nernst Heat Theorem could be deduced from the First and Second Laws. Einstein proved him wrong. What can be inferred from the First and Second Laws is that the heat capacity at T = 0 is zero.

Consider the relation $\Delta G = \Delta H - T\Delta S$. As $T \to 0$, $\Delta G \approx \Delta H$ as shown in Fig. 6.2 (provided ΔS is finite). However, this does not mean $\Delta S(0) = 0$. For ΔS to be zero, the slope $[\partial \Delta G/\partial T]_P$ must be zero. The Second Law does not require it.

We can draw some conclusions from the First and Second Laws

$$(\partial \Delta G/\partial T)_P = -\Delta S = (\Delta G - \Delta H)/T \qquad (6\text{-}5)$$

$$\lim_{T\to 0}(\partial \Delta G/\partial T)_P = \lim_{T\to 0}(\Delta G - \Delta H)/T \approx 0/0 \qquad (6\text{-}6)$$

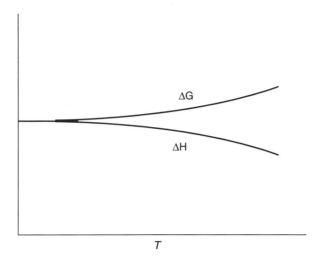

Figure 6.2   Schematic representation of the variation of ΔG and ΔH with temperature.

Differentiating numerator and denominator with respect to T (l'Hospital's Rule) shows that

$$\lim_{T\to0}(\partial\Delta G/\partial T)_P = \lim_{T\to0}[(\partial\Delta G/\partial T)_P - (\partial\Delta H/\partial T)_P]/1 \qquad (6\text{-}7a)$$

and thus

$$\lim_{T\to0}(\partial\Delta H/\partial T)_P = \lim_{T\to0}\Delta C_P = 0 \qquad (6\text{-}7b)$$

The same can be proved for C_P, namely, $\lim_{T\to0}C_P = 0$.

CHAPTER 7

GENERAL CONDITIONS FOR EQUILIBRIUM AND STABILITY

A system is in equilibrium when its properties and the properties of the surroundings do not change with time. If the surrounding properties change, then the process is one of *steady state*.

In mechanics, one can distinguish between four kinds of equilibriums: 1) stable, 2) metastable, 3) neutral, and 4) unstable (see Figure 7.1, in which V represents the potential and r the displacement).

In thermodynamics, there are also stable and metastable equilibriums but not unstable equilibriums:

- *Stable Equilibrium*: ice at 263 K and 1 atm
- *Metastable Equilibrium*: supercooled water at 263 K and 1 atm
- *Neutral Equilibrium*: ice and water at 273 K and 1 atm

In mechanics, the conditions for equilibrium require that $dV/dr = 0$ for all types of equilibrium. For stable and metastable equilibriums, we must also have $d^2V/d^2r > 0$; for neutral equilibrium, $d^2V/d^2r = 0$; and for unstable equilibrium, $d^2V/d^2r < 0$.

In thermodynamics, the inequalities of the thermodynamic functions provide a clue as to the presence or absence of equilibrium and stability conditions. Specifically, for an isolated system *not in equilibrium*,

Thermodynamics and Introductory Statistical Mechanics, by Bruno Linder
ISBN 0-471-47459-2 © 2004 John Wiley & Sons, Inc.

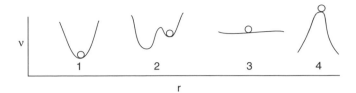

Figure 7.1   Illustration of various kinds of equilibria in mechanics.

$\Delta S_{isolated} > 0$. A plot of $S_{isolated}$ vs. time would look like the diagram in Figure 7.2a. A plot of S vs. some thermodynamic quantity, such as the progress variable, ξ, (Section 9.5) may look like the diagram in Figure 7.2b. The equilibrium value is the maximum value and can be reached from either side. Conversely, if a system is in equilibrium to start with, then any variation that takes it away from equilibrium will result in a decrease in entropy. This can be handled mathematically by introducing the concept of "virtual" variation.

7.1   VIRTUAL VARIATIONS

A *virtual variation* is a variation that takes a system in equilibrium (under a set of constraints) away from it. This can be accomplished, in principle, by adding more constraints. A virtual variation does not mean that the system moves from an equilibrium state to a state of nonequilibrium. Rather, it

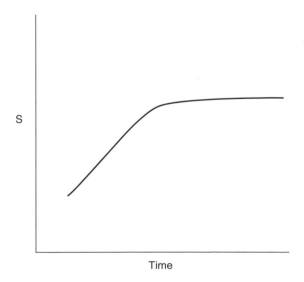

Figure 7.2a   Schematic representation of the variation of S with time.

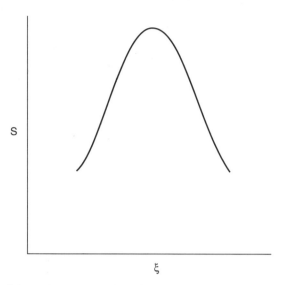

Figure 7.2b   Schematic representation of the variation of S with progress variable.

moves away from one equilibrium state to another (less favorable) equilibrium state. A virtual variation will cause the entropy, which has a maximum at equilibrium, to decrease and a thermodynamic potential (E, H, A, or G), which has a minimum at equilibrium, to increase.

Let us denote a change in the system due to a virtual variation by the symbol Д to distinguish it from the ordinary difference symbol, Δ. Thus, for example, ДS stands for $S - S_{equil}$ or $S - S_0$. [Here, the subscript zero (0) stands for equilibrium, not absolute zero.] To use the idea of virtual variation as applied to S, we expand ДS in a Taylor series in terms of differentials of 1st, 2nd, etc., order:

$$ДS = \delta^{(1)}S + \delta^{(2)}S + \delta^{(3)}S + \cdots < 0 \qquad (7\text{-}1)$$

From the first-order differential, we can obtain conditions for equilibrium, from the second-order differential, we can obtain conditions for stability.

Working with entropy is awkward because of the temperature in the denominators:

$$dS = dE/T + (P/T)dV - \Sigma_i(\mu_i/T)dn_i \qquad (7\text{-}2)$$

which has to be differentiated. It is a lot easier to work with the other thermodynamic potentials E, H, A, and G. The most general results are obtained from E, which we will use for illustrative purposes.

7.2 THERMODYNAMIC POTENTIALS—INEQUALITIES

We have already shown (in Section 5.4) that for a closed system in the absence of "other" work, $\Delta A_{T,V} \leq 0$ (Eq. 5-21b) and $\Delta G_{T,P} \leq 0$, (Eq. 5.23b) where the equal sign refers to a reversible change and the unequal sign to an irreversible transformation. We now show that $\Delta E_{S,V} \leq 0$ and $\Delta H_{S,P} \leq 0$.

Because $dE = dq + dw$ and $dq \leq TdS$, it follows that $dE \leq TdS - PdV$ and, if only PV-work is present,

$$\Delta E_{S,V} \leq 0 \qquad\qquad (7\text{-}3)$$

Also,

$$dH = TdS + VdP \quad \text{and} \quad \text{if } dq \leq TdS,$$

$$\Delta H_{S,P} \leq 0 \qquad\qquad (7\text{-}4)$$

If extended to open systems, we have (Eqs. 5-51 and 5-52)

$$dE = dq + dw + \Sigma_i(\overline{E}_i + P\overline{V}_i)dn_i \qquad\qquad (7\text{-}5)$$

$$dS = dq_{rev}/T + \Sigma_i \overline{S}_i dn_i \qquad\qquad (7\text{-}6)$$

or, in general

$$dq \leq TdS - T\Sigma_i \overline{S}_i dn_i \qquad\qquad (7\text{-}7)$$

Accordingly,

$$dE \leq TdS + dw + \Sigma_i(\overline{E}_i + P\overline{V}_i - T\overline{S}_i)dn_i \qquad\qquad (7\text{-}8)$$

$$dE \leq TdS + dw + \Sigma_i \mu_i dn_i \qquad\qquad (7\text{-}9)$$

and thus, if no other work than PV-work is present, $\Delta E_{S,V;\,\underline{n}_i} \leq 0$.

In a virtual variation, conditions for *stable equilibrium* require that

$$Д E_{S,V;\underline{n}_i} > 0 \qquad\qquad (7\text{-}10)$$

The subscript $\underline{n}_i$ serves to indicate that all n_i are held constant.

Similar arguments can be used to generalize the other thermodynamic potential functions, resulting in the requirements for stable equilibrium

$$ДH_{S,P;\,\underline{n}_i} = H - H_0 > 0 \qquad (7\text{-}11a)$$

$$ДA_{T,V;\,\underline{n}_i} = A - A_0 > 0 \qquad (7\text{-}11b)$$

$$ДG_{T,P;\,\underline{n}_i} = G - G_0 > 0 \qquad (7\text{-}11c)$$

In each case, the subscript 0 refers to the equilibrium value of the function. First, let us consider the stability conditions from the energy expression. Let us expand ДE (without restrictions) in a Taylor series.

$$
\begin{aligned}
ДE &= E - E_0 \\
&= (\partial E/\partial S)_{V;\,\underline{n}_i}(S - S_0) + (\partial E/\partial V)_{S;\,\underline{n}_i}(V - V_0) \\
&\quad + \Sigma_i(\partial E/\partial n_i)_{S,V;n_{j\neq i}}(n_i - n_{i0}) \\
&\quad + 1/2\{(\partial^2 E/\partial S^2)_{V;\,\underline{n}_i}(S - S_0)^2 + 2(\partial^2 E/\partial V\partial S)(S - S_0)(V - V_0) \\
&\quad + (\partial^2 E/\partial V^2)_{S;\,\underline{n}_i}(V - V_0)^2 + \Sigma_i\partial^2 E/\partial n_i^2)_{n_{j\neq i}}(n_i - n_{i0})^2\} + \cdots + \text{other cross terms}
\end{aligned}
$$

$$(7\text{-}12)$$

which we write as a sum of variations ($\delta S = S - S_0, \delta V = V - V_0$ etc.)

$$ДE = T\delta S - P\delta V + \Sigma_i\mu_i\delta n_i + \text{terms in}(\delta S)^2, (\delta V)^2, (\delta n_i)^2, \text{etc.} \quad (7\text{-}13a)$$

For short,

$$ДE = \delta^{(1)}E_{S,V;\underline{n}_i} + \delta^{(2)}E_{S,V;\,\underline{n}_i} + \delta^{(3)}E_{S,V;\,\underline{n}_i} + \text{cross terms in } \delta S\delta V, \text{etc.}$$

$$(7\text{-}13b)$$

The requirement that $ДE_{S,V;\,\underline{n}_i} > 0$ can be satisfied in many ways. For some systems, the first-order variation is already positive, $\delta^{(1)}E > 0$, and $ДE_{S,V;\,n}$ is automatically satisfied. For other systems and in fact for most, the first-order differential is zero, i.e., $\delta^{(1)}E = 0$, and the higher-order differentials must be examined. (We shall refer to such systems as "normal.") Sometimes, $\delta^{(2)}E = 0$, then third- and fourth-order differentials must be considered. In summary

Criteria for equilibrium : $\delta^{(1)}E_{S,V;\,\underline{n}_i} \geq 0$

Criteria for stability : $\delta^{(2)}E_{S,V;\,\underline{n}_i} \geq 0$

Criteria for higher-order stabilities : $\delta^{(n)}E_{S,V;\,\underline{n}_i} > 0$ (when $\delta^{(2)})E_{S,V;\,\underline{n}_i} = 0$).

$$(7\text{-}14)$$

7.3 EQUILIBRIUM CONDITION FROM ENERGY

Let us consider a system that we imagine is divided into two parts. The parts may be real phases or they may be parts separated by an imaginary boundary. Denote these phases as 1 and 2. Variations in S, V, or n may occur in each "phase," but the total values of S, V, and n must remain constant, as required by stability conditions. Thus

$$\delta^{(1)}E_{S,V;\,\underline{n}_i} = T^{(1)}\delta S^{(1)} + T^{(2)}\delta S^{(2)} - P^{(1)}\delta V^{(1)} - P^{(2)}\delta V^{(2)}$$

$$+ \Sigma_i \mu_i^{(1)}\delta n_i^{(1)} + \Sigma_i \mu_i^{(2)}{}_i\delta n_i^{(2)} \geq 0 \qquad (7\text{-}15)$$

Here, the superscripts refer to the phases or parts of the system. Recall that the T, P, and μ values are derivatives of E evaluated at equilibrium. Thus, the values of intensive variables are the equilibrium values from which the virtual variations have occurred. Furthermore, it should be clear that, because the total S, V, and n are fixed, the sums of their variations in the two phases must add up to zero., i.e.

$$\delta S = \delta S^{(1)} + \delta S^{(2)} = 0$$

$$\delta V = \delta V^{(1)} + \delta V^{(2)} = 0 \qquad (7\text{-}16)$$

$$\delta n_i = \delta n_i^{(1)} + \delta n_i^{(2)} = 0$$

Examples of these are given in the following sections.

7.3.1 Boundary Fully Heat Conducting, Deformable, Permeable (Normal System)

Suppose all variations are zero except the fluctuations in entropy; that is, $\delta S^{(1)}$ or $\delta S^{(2)}$ are not zero. All other variations are zero. Obviously, $\delta S^{(1)} = -\delta S^{(2)}$.

It follows from Eq. 7-15 that

$$\delta^{(1)}E = [T^{(1)} - T^{(2)}]\delta S^{(1)} \geq 0 \qquad (7\text{-}17)$$

$$\text{if } \delta S^{(1)} > 0, \text{then } T^{(1)} \geq T^{(2)} \qquad (7\text{-}18a)$$

$$\text{if } \delta S^{(1)} < 0, \text{then } T^{(1)} \leq T^{(2)} \qquad (7\text{-}18b)$$

Obviously, the inequalities contradict each other (the T terms refer to the initial equilibrium T values) and the only valid relation is

$$T^{(1)} = T^{(2)} = T \qquad (7\text{-}18c)$$

EXERCISES

1. Let there be fluctuations only in the volume. Show that

$$\text{if } \delta V^{(1)} > 0 \text{ then } P^{(2)} \geq P^{(1)} \qquad (7\text{-}19a)$$

$$\text{if } \delta V^{(1)} < 0 \text{ then } P^{(2)} \leq P^{(1)} \qquad (7\text{-}19b)$$

and so,

$$P^{(1)} = P^{(2)} = P \qquad (7\text{-}19c)$$

2. Show that, if fluctuations are only allowed in n_i,

$$\mu_i^{(1)} = \mu_i^{(2)} = \mu_i \qquad (7\text{-}20)$$

In summary, for systems in which temperature, pressure, and chemical potentials are uniform throughout ("normal" systems), $\delta^{(1)}E_{S,V;\,\underline{n}_i} = 0$.

7.3.2 Special Cases: Boundary Semi-Heat Conducting, Semi-Deformable, or Semi-Permeable

We can apply the same procedure to cases in which the partition is semi-heat conducting, semi-deformable, or semi-permeable. For example, partition is semi-heat conducting, so that heat can only flow from 2 to 1, in which case, $\delta S^{(1)} > 0$. Again, holding all variations fixed except $\delta S^{(1)} = -\delta S^{(2)}$, we get

$$[T^{(1)} - T^{(2)}]\delta S^{(1)} \geq 0 \qquad (7\text{-}21)$$

or $T^{(1)} \geq T^{(2)}$. In other words, the system can be in equilibrium if $T^{(1)}$ equals $T^{(2)}$ or is greater than $T^{(2)}$, but $T^{(2)}$ cannot be less that $T^{(1)}$.

EXERCISES

3. Suppose the partition is semi-deformable, in such a way that only $V^{(1)}$ can increase [i.e., $\delta V^{(1)} = -\delta V^{(2)} \geq 0$]. All other variations are zero. Show that

$$P^{(2)} \geq P^{(1)} \qquad (7\text{-}22)$$

4. Suppose the partition is semi-permeable to species i in such a way that i can only flow from 2 to 1. Show that

$$\mu_i^{(1)} \geq \mu_i^{(2)} \qquad (7\text{-}23)$$

Note: If the partition is impermeable to all kinds of variations, then each phase can have arbitrary temperature, pressure, and chemical potential values.

7.4 EQUILIBRIUM CONDITIONS FROM OTHER POTENTIALS

The foregoing analysis based on $\delta^{(1)}E_{S,V;\, n_i}$ can be applied to the other thermodynamic potentials, namely, $\delta^{(1)}H_{S,P;\, n_i}$, $\delta^{(1)}A_{T,V;\, n_i}$, and $\delta^{(1)}G_{T,P;\, n_i}$. Obviously, we must pay attention to the constraints indicated by the subscripts. Thus, when an intensive variable is held constant, we cannot divide the system into two parts, assuming that T or P decreases in one part and increases by the same amount in the other part. The intensive variables are not additive. The constraints allow virtual variations between the "phases" of the extensive variables only; the intensive variables must be uniform and constant throughout. Thus, from $\delta^{(1)}H_{S,P;\, n_i}$ we can obtain information about the equilibrium conditions of T and μ_i but not P. However, we already know the condition for P: it must be uniform throughout. Similarly, from $^{(1)}\delta A_{T,V;\, n_i}$ we can get the equilibrium criteria for P and μ_i but not for T. From $\delta^{(1)}G_{T,P;\, n_i}$, we can only obtain equilibrium conditions for μ_i.

7.5   GENERAL CONDITIONS FOR STABILITY

In the previous sections, we developed criteria for stable equilibrium for a number of systems and for different thermodynamic potentials. In this section, we will discuss the nature of the equilibrium states, namely, whether the systems are in stable or neutral (or undetermined) equilibrium or whether they are unstable. In thermodynamics, we do not have *lasting* unstable equilibria. Using the symbol $>$ to represent stable equilibrium and the sign $=$ to represent neutral equilibrium, we may write

$$\text{ДE}_{S,V;\,\underline{n}_i} \geq 0 \qquad\qquad (7\text{-}24)$$

If we use the other potentials, the conditions for stability are

$$\text{ДH}_{S,P;\,\underline{n}_i} \geq 0 \qquad\qquad (7\text{-}25a)$$

$$\text{ДA}_{T,V;\,\underline{n}_i} \geq 0 \qquad\qquad (7\text{-}25b)$$

$$\text{ДG}_{T,P;\,\underline{n}_i} \geq 0 \qquad\qquad (7\text{-}25c)$$

7.6   STABILITY CONDITIONS FROM E

The equilibrium criteria, as shown in preceding sections, require that for "normal" systems T, P, and μ_i be constant and $\delta^{(1)}E_{S,V;\,\underline{n}_i} = 0$. To satisfy the stability conditions, one must consider the second-order (or higher-order) variations. As shown before

$$\delta^{(2)}E = 1/2[\partial^2 E/\partial S^2)_{V;\,\underline{n}_i}(S - S_0)^2 + (\partial^2 E/\partial V^2)_{S;\,\underline{n}_i}(V - V_0)^2$$
$$+ \sum_i (\partial^2 E/\partial n_i^2)_{S,V;\,\underline{n}_{j\neq i}}(n - n_i)^2 + \text{cross terms}] \qquad (7\text{-}26a)$$

$$= 1/2[(T/C_V)(\delta S)^2 - (\partial P/\partial V)_{S;\,\underline{n}_i}(\delta V)^2$$
$$+ \sum_i (\partial\mu_i/\partial n_i)_{S,V;n_{j\neq i}}(\delta n_i)^2 + \text{cross terms}] \qquad (7\text{-}26b)$$

Let us again divide the system into two parts ("phases"), allowing fluctuations to occur between the parts but keeping in mind that the total S, V, and n_i must be constant. Because $\delta S^{(1)} = -\delta S^{(2)}$ and so $(\delta S^{(1)})^2 = (-\delta S^{(2)})^2 = (\delta S^{(2)})^2$, and so forth, and for normal systems $T^{(1)} = T^{(2)} = T, P^{(1)} = P^{(2)} = P$ and $\mu_i^{(1)} = \mu_i^{(2)} = \mu_i$, we obtain for stable equilibrium

$$\delta^2 E_{S,V;\, \underline{n}_i} = 1/2\{[T/C_V^{(1)} + T/C_V^{(2)}](\delta S^{(1)})^2$$

$$+ [-(\partial P/\partial V^{(1)})_{S^{(1)},\, \underline{n}_i^{(1)}} - (\partial P/\partial V^{(2)})_{S^{(2)},\, \underline{n}_i^{(2)}}](\delta V^{(1)})^2$$

$$+ \Sigma_i[(\partial \mu_i/\partial n_i^{(1)})_{S^{(1)},V^{(1)},\, \underline{n}_{j\neq i}^{(1)}}, +(\partial \mu_i/\partial n_i)_{S^{(2)},V^{(2)},\, \underline{n}_{j\neq i}^{(2)}}](\delta n_i^{(1)})^2$$

$$+ \text{cross terms in } \delta S^{(1)}, \delta V^{(1)}, \text{etc.}\} > 0 \qquad (7\text{-}27)$$

Holding all variations fixed except $\delta S^{(1)} = -\delta S^{(2)}$ we get

$$T[1/C_V^{(1)} + 1/C_V^{(2)}](\delta S^{(1)})^2 > 0 \qquad (7\text{-}28)$$

Because $T > 0$ except at absolute zero, the quantity within brackets must be positive. If 1 and 2 are portions of the same homogeneous phase, we can denote the heat capacities as $C_V^{(1)} = n^{(1)}\overline{C}_V$ and $C_V^{(2)} = n^{(2)}\overline{C}_V$. Obviously, $\overline{C}_V > 0$ when $T > 0$. When $T = 0$, $\overline{C}_V = 0$, as we already saw (Section 6.2), but $T/C_V > 0$. We can express these results by writing

$$C_V \geq 0 \qquad (7\text{-}29)$$

where the $=$ sign refers to 0 K and $>$ refers to a finite temperature.

Holding all variations fixed except $\delta V^{(1)} = -\delta V^{(2)}$, we get

$$-[\partial P/\partial V^{(1)})_{S^{(1)},\, \underline{n}_i^{(1)}} + (\partial P/\partial V^{(2)})_{S^{(2)},\, \underline{n}_i^{(2)}}(\delta V^{(1)})^2 > 0$$

or,

$$[(\partial P/\partial V^{(1)}_{S^{(1)},\, \underline{n}_i^{(1)}} + (\partial P/\partial V^{(2)})_{S^{(2)},\, \underline{n}_i^{(2)}}] < 0 \qquad (7\text{-}30)$$

Again, for a homogeneous system, we must have

$$(\partial P/\partial V)_{S;\, \underline{n}_i} < 0 \qquad (7\text{-}31a)$$

or

$$\kappa = -1/V(\partial V/\partial P)_{S;\, \underline{n}_i} > 0 \qquad (7\text{-}31b)$$

Note: This condition holds for internal stability of a homogeneous system; it does not apply to the critical point. For this, the variation of $\delta^2 E$ with respect to V is zero, and one must go to higher-order differentials, namely, $\delta^4 E$ (it can be shown that if $\delta^2 E = 0$ then $\delta^3 E = 0$), to obtain the conditions for stability.

Holding all variations fixed except $\delta n_i^{(1)} = -\delta n_i^{(2)}$, gives

$$[(\partial\mu_i/\partial n_i^{(1)})_{S^{(1)},V^{(1)},n_{j\neq i}^{(1)}} + (\partial\mu_i/\partial n_i^{(2)})_{S^{(2)},V^{(2)},n_{j\neq i}^{(2)}}](\delta n_i)^2 > 0 \qquad (7\text{-}32)$$

This condition is usually expressed in terms of mole fractions. For a two-component system, the mole fractions in phase 1 are

$$x_1^{(1)} = n_1^{(1)}/(n_1^{(1)} + n_2^{(1)}) \quad \text{and} \quad x_2^{(1)} = n_2^{(1)}/(n_1^{(1)} + n_2^{(1)}) \qquad (7\text{-}33)$$

It is easy to show that

$$dn_1^{(1)} = (n^{(1)}/x_2^{(1)})dx_2^{(1)} \qquad (7\text{-}34a)$$

where $n^{(1)} = n_1^{(1)} + n_2^{(1)}$ and similarly, for phase 2

$$dn_1^{(2)} = (n^{(2)}/x_2^{(2)})dx_1^{(2)} \qquad (7\text{-}34b)$$

For species 1

$$(x_2^{(1)}/n^{(1)})(\partial\mu_1/\partial x_1^{(1)})_{S^{(1)},V^{(1)},\underline{n}_2^{(1)}} + (x_2^{(2)}/n^{(2)})(\partial\mu_1/\partial x_1^{(2)})_{S^{(2)},V^{(2)},\underline{n}_2^{(2)}} > 0$$

$$(7\text{-}35)$$

For a homogeneous system,

$$(\partial\mu_i/\partial x_i)_{S,V;\,\underline{n}_i} > 0 \qquad (7\text{-}36)$$

7.7 STABILITY CONDITIONS FROM CROSS TERMS

We have so far not considered conditions arising from cross terms. Cross terms give rise to additional criteria, which can be useful. For example, the stability conditions for fixed n_i, which include cross terms, may be written concisely

$$\delta^2 E_{S,V;n_i} = 1/2[(\partial^2 E/\partial S^2)_V(\delta S)^2 + 2(\partial^2 E/\partial S\partial V)\delta S\delta V$$
$$+ (\partial^2 E/\partial V^2)_S(\delta V)^2] > 0 \qquad (7\text{-}37)$$

The quantity within the brackets must be positive for any conceivable variation in S and V. (Such functions are called positive definitive.) It can be shown that, for a function to be positive definitive, all roots, λ, of

Eq. 7-38 must be greater than zero

$$\begin{vmatrix} \partial^2 E/\partial S^2 - \lambda & \partial^2 E/\partial S \partial V \\ \partial^2 E/\partial V \partial S & \partial^2 E/\partial V^2 - \lambda \end{vmatrix} = 0 \qquad (7\text{-}38)$$

The λ so obtained is

$$\lambda = 1/2(\partial^2 E/\partial S^2) + (\partial^2 E/\partial V^2) \pm 1/2[(\partial^2 E/\partial S^2 + \partial^2 E/\partial V^2)^2$$
$$- 4(\partial^2 E/\partial S^2)(\partial^2 E/\partial V^2) + 4(\partial^2 E/\partial S \partial V)^2]^{1/2} \qquad (7\text{-}39)$$

If λ is to be positive for any conceivable variation of S or V, then

$$-4(\partial^2 E/\partial S^2)(\partial^2 E/\partial V^2) + 4(\partial^2 E/\partial V \partial S)^2 < 0 \qquad (7\text{-}40a)$$

or,

$$(\partial^2 E/\partial V \partial S)^2 < (\partial^2 E/\partial S^2)(\partial^2 E/\partial V^2) \qquad (7\text{-}40b)$$

which shows that

$$(T/C_V)(-\partial P/\partial V)_S > [(\partial T/\partial V)_S]^2 \qquad (7\text{-}40c)$$

7.8 STABILITY CONDITIONS FROM OTHER POTENTIALS

Up to now, we have only considered stability conditions derived from $\delta^2 E_{S,V;\,\underline{n}_i}$, which for "normal" systems in stable equilibrium gave $C_V \geq 0$, $(\partial P/\partial V)_{S;n_i} < 0$, and $(\partial \mu_i/\partial n_i)_{S,V;\,n_{j\neq i}} > 0$. Had we used

i) $\delta^2 H_{S,P;\,\underline{n}_i}$, we would have obtained $C_P \geq 0$ and $(\partial \mu_i/\partial n_i)_{S,P;\,n_{j\neq i}} > 0$

$$(7\text{-}41)$$

ii) $\delta^2 A_{T,V;\underline{n}_i}$, we would have obtained $(\partial/P!\partial V)_{T,V;\,\underline{n}_i} < 0$

and $(\partial \mu_i/\partial n_i)_{T,V;\,n_{j\neq i}} > 0$ $\qquad (7\text{-}42)$

iii) $\delta^2 G_{T,P;n}$ we would have obtained $(\partial \mu_i/\partial n_i)_{T,P;\,n_{j\neq i}} > 0$. $\qquad (7\text{-}43)$

7.9 DERIVATIVES OF THERMODYNAMIC POTENTIALS WITH RESPECT TO INTENSIVE VARIABLES

In the case of "normal" systems (fully heat-conducting, -deformable, -permeable), the foregoing analyses allowed variations in the extensive variable S, V, and n_i but not in the intensive variables T, P, and μ_i. What can be said about the derivatives of the thermodynamic potentials with respect to the intensive variables? Obviously, the technique devised to obtain the derivatives of the potential functions with respect to the extensive variables cannot be used for intensive variables, as noted before. It makes no sense to say that, for example, the sum of the temperatures of the two parts of the system adds up to the total temperature. Nonetheless, there are other ways to establish the conditions for stable equilibrium of the derivatives of the thermodynamic potentials with respect to the intensive variables, simply by deducing them from the conditions of the potentials with respect to the extensive variables.

In particular, from the condition of stable equilibrium based on $(\partial^2 E/\partial V^2)_{S;\,\underline{n}_i} > 0$, we found that $(\partial P/\partial V)_{S;\,\underline{n}_i} < 0$.

1) Since, $(\partial^2 H/\partial P^2)_{S;\,\underline{n}_i} = (\partial V/\partial P)_{S;\,\underline{n}_i} = 1/(\partial P/\partial V)_{S;\,\underline{n}_i}$, we obtain

$$(\partial^2 H/\partial P^2)_{S;\,\underline{n}_i} < 0 \qquad (7\text{-}44a)$$

2) Similarly, from $(\partial^2 E/\partial S^2)_{V;\,\underline{n}_i} > 0$, we deduce that

$$(\partial^2 A/\partial T^2)_{V;\,\underline{n}_i} = -C_V/T < 0 \qquad (7\text{-}44b)$$

3) From $(\partial^2 H/\partial S^2)_{P;\,\underline{n}_i}$, we obtain

$$(\partial^2 G/\partial T^2)_{P;\,\underline{n}_i} = -C_{P,n}/T < 0 \qquad (7\text{-}44c)$$

4) From $(\partial^2 A/\partial V^2)_{T;\,\underline{n}_i} = -(\partial P/\partial V)_{T;\,\underline{n}_i}$, we get

$$(\partial^2 G/\partial P^2) = (\partial V/\partial P)_{T;n} < 0 \qquad (7\text{-}44d)$$

These results may be summarized as follows: *All second derivatives of the thermodynamic potentials (E,H,A,G) are* **convex** *functions of the extensive variables and* **concave** *functions of the intensive variables.*

CHAPTER 8

APPLICATION OF THERMODYNAMICS TO GASES, LIQUIDS, AND SOLIDS

8.1 GASES

An *equation of state* is an equation that connects the intensive variables T, P, $\overline{V}$, $x_1, x_2, \ldots, x_{r-1}$. The symbol $\overline{V}$ stands for the volume per unit mass, and x_i stands for the mole fraction of species i. Every system in thermal equilibrium possesses an equation of state. Equations of state must be determined experimentally: thermodynamics cannot predict them.

Experimentally, all gases have these properties:

1) As $P \to 0$, $\lim P\overline{V}/RT = 1$; that is, the gas behaves as an ideal gas.
2) The gases undergo phase transitions and have a critical point; that is $(\partial P/\partial V)_T = 0$ and $(\partial^2 P/\partial V^2)_T = 0$
3) No exact analytical expression exists for an equation of state, but approximate relations are available, representing *ideal gas (Eq. 8-1), elastic hard sphere (Eq. 8-2)*

$$PV = nRT \tag{8-1}$$

$$P(V - b) = nRT, \tag{8-2}$$

Thermodynamics and Introductory Statistical Mechanics, by Bruno Linder
ISBN 0-471-47459-2 © 2004 John Wiley & Sons, Inc.

where b is a constant, and *Van der Waals equation (Eq. 8-3)*

$$(P + a/V^2)(V - b) = nRT, \qquad (8\text{-}3)$$

where a and b are constants.

The van der Waals equation violates the stability conditions in some regions (as will be shown later) but predicts the existence of a critical point. Denoting the critical volume, critical pressure and critical temperature, respectively, as V_c, P_c and T_c, one can show that

$$V_c = 3b, \quad P_c = a/27b^2 \quad \text{and} \quad T_c = 8a/27Rb \qquad (8\text{-}4)$$

There are other more complicated equations of state (such as the *Beattie-Bridgeman* and the *Berthelot* equation). These are frequently discussed in physical chemistry books. We will discuss two other equations of state (the Law of Corresponding States and the Virial Equation of State) below.

For the *Law of Corresponding States*, many gases, especially those with parameters like the van der Waals a and the van der Waals b (in addition to R), can be fitted into a universal equation of state in terms of *reduced temperature* ($\tau = T/T_c$), *reduced pressure* ($\pi = P/P_c$), *and reduced volume* ($\phi = V/V_c$). The Corresponding Law equation of state for a van der Waals gas is

$$\pi = 8\tau/(3\phi - 1) - 3/\phi^2 \qquad (8\text{-}5)$$

Note that there are no parameters characteristic of the substance.

There are two forms of the *Virial Equation of State*:

$$P\overline{V} = RT[1 + A_2(T)P + A_3(T)P^2 + \cdots] \qquad (8\text{-}6)$$

$$P\overline{V} = RT[1 + B_2(T)/\overline{V} + B_3(T)/\overline{V}^2 + \cdots] \qquad (8\text{-}7)$$

When solving problems involving enthalpy relations, it is more convenient to use the virial form in terms of P. However, the values for the virial coefficients are usually given in term of the B and not the A coefficients. To relate the A and B coefficients, write P in terms of $\overline{V}$, using *Equation 8-7* and then substitute those P values in *Equation 8-6*. Comparing the coefficients of $1/\overline{V}$ gives

$$P\overline{V} = RT + B_2(T)P + \cdots \qquad (8\text{-}8)$$

Let us consider an example. The van der Waals equation of state (per mole) in virial form is

$$(P + a/\overline{V}^2)(\overline{V} - b) = RT$$

or

$$P\overline{V} = RT\overline{V}/(\overline{V} - b) - a/\overline{V} \qquad (8\text{-}9)$$

Expanding the right-hand side of *Eq. 8-9* gives

$$P\overline{V} = RT[1 + b/\overline{V} + b^2\overline{V}^2 +] - a/\overline{V}$$
$$= RT[1 + (b - a/RT)1/\overline{V} + \cdots] \qquad (8\text{-}10)$$

Thus

$$B_2 = (b - a/RT) \qquad (8\text{-}11)$$

Also, using *Eq. (8-8)* shows that $\lim_{P\to 0}[\partial(P\overline{V})/\partial P]_T = B_2(T)$

8.2 ENTHALPY, ENTROPY, CHEMICAL POTENTIAL, FUGACITY

8.2.1 Enthalpy

From the general relation

$$dH = (\partial H/\partial T)_P dT + (\partial H/\partial P)_T dP \qquad (8\text{-}12)$$

we get, on integration between the limits 1 and 2:

$$H(T_2, P_2) - H(T_1, P_1) = \int_1^2 (\partial H/\partial T)_P \, dT + \int_1^2 (\partial H/\partial P)_T dP. \qquad (8\text{-}13)$$

This expression is not very useful because $(\partial H/\partial p)_T$ has to be known over the entire range of temperatures and pressures. The integral over pressure causes no problems, even if the lower limit is extended to $P = 0$, because at zero pressure the gas behaves ideally. However, the temperature derivative is generally not known over the full range of temperatures. To get around this difficulty, it is common practice to treat the integral over T as an indefinite integral and the integral over P as a definite integral. Referring to molar quantities, we write

$$\overline{H}(T, P) = \int^{T} (\partial\overline{H}/\partial T)_{P\to 0} dT + \overline{H}_0 + \int_{0}^{P} (\partial\overline{H}/\partial P')_T dP' \qquad (8\text{-}14a)$$

$$= \int^{T} \overline{C}_{P\to 0}(T)\, dT + \overline{H}_0 + \int_{0}^{P} [\overline{V} - T(\partial\overline{V}/\partial T)_{P'}]\, dP' \qquad (8\text{-}14b)$$

$$= \overline{H}^{0}(T) + \int_{0}^{P} [\overline{V} - T(\partial\overline{V}/\partial T)_P]\, dP' \qquad (8\text{-}14c)$$

where $\overline{H}_0(T)$ is an integration constant, and $\overline{H}^{0}(T)$ is the standard enthalpy, equal to the first two terms on the right of Eq. 14-b.

Note: As $P \to 0$, the gas behaves as an ideal gas, and E and H depend only on the temperature (as noted before), as does $C_{P\to 0}$. It is clear that for an ideal gas the second term is zero and thus,

$$\overline{H}(T, P) = \overline{H}^{\text{ideal gas}}(T, P) = \overline{H}^{0}(T) \qquad (8\text{-}15)$$

Thus, $\overline{H}^{0}$ (T,P) is the enthalpy of the real gas as $P \to 0$.

We present an application of the above discussion. For a van der Waals gas (in virial form), using Eq. 8.8

$$\overline{H}(T, P) - \overline{H}^{0}(T) = \int_{0}^{P} dP'(RT/P' + B_2 - TR/P' - TdB_2/dT) \qquad (8\text{-}16a)$$

$$= (B_2 - T\, dB_2/dT)P \qquad (8\text{-}16b)$$

$$= (b - 2a/RT)P \qquad (8\text{-}16c)$$

8.2.2 Entropy

The problem for entropy is similar to that of enthalpy, but there are additional complications. The differential form of S (see Eq. 5. 13a) is

$$d\overline{S}(T, P) = (\partial\overline{S}/\partial T)_P dT + (\partial\overline{S}/\partial P)_T dP \qquad (8\text{-}17a)$$

$$= (\overline{C}_P/T)dT - (\partial\overline{V}/\partial T)_P dP \qquad (8\text{-}17b)$$

The last term diverges ("blows up") as $P \to 0$, since the gas approaches ideal gas behavior and $(\partial\overline{V}/\partial T)_P \to R/P \to \infty$ as $P \to 0$. We can get around this difficulty by adding to both sides of the equation $RdP/P = R\, d\ln(P/P^0)$

(we have included the unit pressure, P^0, to make the logarithmic argument dimensionless). *Equation 8-17* then becomes

$$d[\overline{S}(T,P) + R \ln(P/P^0)] = [C_{P\to0}(T)/T]dT + [R/P - (\partial\overline{V}/\partial T)]dP \quad (8\text{-}18)$$

Integrating dT within indefinite limits and dP within definite limits, yields

$$[\overline{S}(T,P)] + R \ln(P/P^0) = \int^T [\overline{C}_{P\to0}(T)/T]dT + \overline{S}_0 + \int_0^P [R/P' - (\partial\overline{V}/\partial T)_P]dP'$$

$$(8\text{-}19)$$

$$\overline{S}(T,P) = \overline{S}^0(T) - R \ln(P/P^0) + \int_0^P [R/P' - (\partial\overline{V}/\partial T)_{P'}]dP' \quad (8\text{-}20)$$

where $\overline{S}_0$ is the integrating constant, and $\overline{S}^0$ is the standard entropy.

For an ideal gas,

$$\overline{S}^{\text{ideal gas}}(T,P) = \overline{S}^0(T) - R \ln(P/P^0) \quad (8\text{-}21)$$

which shows that the standard entropy $\overline{S}^0$ is not equal to the entropy of an ideal gas, except when $P = P^0$. Rather, $\overline{S}^0$ is the entropy of a "hypothetically" ideal gas at unit pressure. This is in contrast to the standard enthalpy, $\overline{H}^0$, which is the enthalpy of the ideal gas without any restrictions.

EXERCISE

Show that the virial entropy expression is

$$\overline{S}(T,P) = \overline{S}^0(T) - R \ln P/P^0 - (dB_2/dT)P + \cdots \quad (8\text{-}22)$$

8.2.3 Chemical Potential

The chemical potential (or molar Gibbs free energy) can now readily be obtained by combining $\overline{H}$ and $\overline{S}$

$$\mu = \overline{H}(T,P) - T\overline{S}(T,P)$$

$$= RT \ln(P/P^0) + \int_0^P [\overline{V} - RT/P']dP'$$

$$+ \int^T \overline{C}_{P \to 0}(T)(1 - T/T^0)dT + [\overline{H}_0 - T S_0]$$

$$= \mu^0(T) + RT \ln(P/P^0) + \int_0^P [\overline{V} - RT/P']dP' \qquad (8\text{-}23)$$

where μ^0 is the standard chemical potential.

For an ideal gas, $\mu^{ideal\ gas}(T,P) = \mu^0(T)$ provided $P = P^0$. Thus, $\mu^0(T)$ is not the chemical potential of an ideal gas but the chemical potential of a "hypothetical" ideal gas at unit pressure.

EXERCISE

Derive expressions for $\mu(T, P) - \mu^0(T)$

1) For a gas obeying the virial equation of state
2) For a gas obeying the van der Waals Equation of State

8.2.4 Fugacity

For an ideal gas, the chemical potential has the form

$$\mu(T, P) = \mu^0(T) + RT \ln(P/P^0) \qquad (8\text{-}24)$$

This is obviously a very simple expression. Is it possible to write an expression for the chemical potential of a real gas having a form as simple as *Equation 8-24*? The answer is yes. This was first done by G. N. Lewis, who introduced the concept of "fugacity," denoted by the symbol f. The fugacity may be defined as

$$f = P \exp\left\{ (1/RT) \int_0^P (\overline{V} - RT/P)dP' \right\} \qquad (8\text{-}25)$$

and so,

$$\ln (f/P^0) = \ln (P/P^0) + 1/RT \int_0^P (\overline{V} - RT/P')dP' \qquad (8\text{-}26)$$

Comparing *Eq. 8-26* with *Eq. 8-23* shows that

$$\mu(T,P) - \mu^0(T) = RT \ln(f/P^0) \qquad (8\text{-}27)$$

which is formally as simple as the ideal gas equation. Thus, the standard chemical potential is the chemical potential of the real gas at unit fugacity. The fugacity can be thought of as an "effective" pressure. But is it useful? According to Guggenheim (1967), "The simplification obtained by the introduction of fugacity is one of appearance of elegance, but leads to nothing quantitative unless we express the fugacity in terms of pressure, and so we are back where we started."

EXERCISE

Derive expressions for the fugacity in terms of

1) the virial equation
2) the van der Waals equation

Note that $\mu(T,P) = \mu^0(T)$ when f (of the real gas) is equal to P^0.

8.3   STANDARD STATES OF GASES

There really is no particular state that can be called standard. One may define standard free energy, standard entropy, and standard enthalpy, but they do not refer to the same state. As noted previously

1) the standard free energy or standard chemical potential, $\mu^0(T)$, is the free energy of the actual gas at unit fugacity, f^0;
2) the standard entropy, $\overline{S}^0(T)$, is the entropy of a hypothetical ideal gas at unit pressure, P^0;
3) the standard enthalpy, $\overline{H}^0(T)$, is the enthalpy of the real gas at zero pressure, $P \to 0$.

This choice must be made for thermodynamic consistency! Thus, if we take $\mu^0(T)$ to be the standard chemical potential of a gas, then by the Gibbs-Helmholtz equation, we have

$$[\partial(\mu^0/T)/\partial(1/T)]_P = \overline{H}^0\,(T) \qquad (8\text{-}28)$$

and

$$(\partial \mu^0 / \partial T)_P = -\overline{S}^0(T) \qquad (8\text{-}29)$$

which are consistent with the definitions of the standard enthalpy and standard entropy of the gas.

Note: Standard free energy and standard enthalpy are associated with particular realizable states (although not the same). The standard entropy does not refer to a real state but a hypothetical one.

8.4 MIXTURES OF GASES

8.4.1 Partial Fugacity

If we write for species i

$$\mu_i(T, P) = \mu_i^0(T) + RT \ln(P_i / P^0) + \int_0^P [\overline{V}_i - RT/P] dP \qquad (8\text{-}30)$$

replace P_i by $x_i P$ and define f_i by

$$f_i = x_i P \exp\left[1/RT \int_0^P (\overline{V}_i - RT/P') dP'\right] \qquad (8\text{-}31)$$

or,

$$\ln(f_i / P^0) = \ln x_i (P/P^0) + 1/RT \int_0^P [\overline{V}i - RT/P'] dP' \qquad (8\text{-}32)$$

we obtain,

$$\mu_i(T, P) = \mu_i^0(T) + RT \ln(f_i / P^0) \qquad (8\text{-}33)$$

8.4.2 Free Energy, Entropy, Enthalpy, and Volume of Mixing of Gases

Let the superscript $\bullet$ denote a pure substance. The free energy of the mixture is represented as $G = \Sigma_i \mu_i n_i$ and that of the pure substances as $G^\bullet = \Sigma_i \mu_i^\bullet n_i$.

The free energy of mixing is defined as

$$\Delta G_{mix} = \Sigma_i\, n_i(\mu_i - \mu_i^{\bullet}) \tag{8-34}$$

Using *Eq. 8-33* and *8-32* gives

$$\Delta G_{mix} = RT\Sigma_i n_i \ln f_i/f_i^{\bullet}$$

$$= RT\Sigma_i\, n_i\, \ln x_i + \Sigma_i n_i \int_0^P (\overline{V}_i - \overline{V}_i^{\bullet})dP' \tag{8-35}$$

$$\Delta S_{mix} = -[\partial\Delta G_{mix}/\partial T]_P$$

$$= -R\Sigma_i n_i \ln x_i - \Sigma_i\, n_i \int_0^P [(\partial\overline{V}_i/\partial T)_{P''} - (\partial\overline{V}_i^{\bullet}/\partial T)_{P'}]dP' \tag{8-36}$$

$$\Delta H_{mix} = \Sigma_i n_i \int_0^P [(\overline{V}_i - \overline{V}_i^{\bullet}) - T(\partial\overline{V}_i/\partial T)_P + T(\partial\overline{V}_i^{\bullet}/\partial T)]dP' \tag{8-37}$$

$$\Delta V_{mix} = \Sigma_i n_i(\overline{V}_i - \overline{V}_i^{\bullet}) \tag{8-38}$$

Note: For a mixture of ideal gases, $\overline{V}_i = \overline{V}_i^{\bullet}$ and

$$\Delta G_{mix} = RT\ \Sigma_i n_i \ln x_i \tag{8-39}$$

$$\Delta S_{mix} = -R\ \Sigma_i n_i \ln x_i \tag{8-40}$$

$$\Delta H_{mix} = 0 \tag{8-41}$$

$$\Delta V_{mix} = 0 \tag{8-42}$$

8.5 THERMODYNAMICS OF CONDENSED SYSTEMS

Consider a single component, one-phase solid or liquid system. (We will take up solutions in Chapter 10.)

There is a sharp contrast between gases on one hand and solids and liquids on the other. In gases (except near the critical point) $\kappa = -\,1/V(\partial V/\partial P)_T$ and of the order of $1/P$. In condensed systems, κ is very much smaller and nearly constant, i.e.,

$$\kappa = -1/V(\partial V/\partial P)_T \approx \text{constant} \tag{8-43}$$

This results in an approximate equation of state, which can be justified by considering

$$dV/V = -\kappa dP \qquad (8\text{-}44)$$

Integrating between $P = 0$ and P for fixed T, yields

$$\int dV/V = -\kappa \int_0^P dP' \qquad (8\text{-}45)$$

Thus,

$$\ln[V(T, P)/V(T, 0)] = -\kappa P \qquad (8\text{-}46)$$

or,

$$V(T, P) \approx V(T, 0)e^{-\kappa P} \approx V(T, 0)(1 - \kappa P) \qquad (8\text{-}47)$$

where V(T,0) is the molar volume extrapolated to $P = 0$. Further simplification is achieved by replacing V(T,0) by V(0,0), which is permissible because $\alpha = 1/V(\partial V/\partial T)_P$ is of the order of $10^{-3}\ T^{-1}$ or even smaller.

8.5.1 The Chemical Potential

Starting with the differential form

$$d\mu = (\partial\mu/\partial T)_P dT + (\partial\mu/\partial P)_T dP \qquad (8\text{-}48a)$$

and using an indefinite limit for the T integration and definite limits for the P integration, we obtain

$$\mu(T, P) = \int^T (\partial\mu/\partial T)_{P\to 0} + \mu_0 + \int_0^P \overline{V}dP' \qquad (8\text{-}48b)$$

$$= \mu^0(T, P = 0) + P\overline{V}(T, 0)(1 - \tfrac{1}{2}\kappa P) \qquad (8\text{-}48c)$$

Note: Variation of μ with pressure is very different for condensed systems than for gases. For solids, κ is of the order of $10^{-6}\ atm^{-1}$; for liquids, it is approximately $10^{-4}\ atm$. In essence, $(1 - 1/2\kappa\mu P) \approx 1$. In other words, $\mu(T,P)$ varies linearly with $PV(T,0)$. (For gases, the variation is logarithmic.) Moreover, for most liquids and solids, $\overline{V}$ ranges between 10 and $100\ cm^3$, so that $P\overline{V} \approx 0.1\ atm \cdot L$. (For gases, $P\overline{V}$ is of the order of RT, i.e., $\approx 20\ atm \cdot L$) Thus, for condensed systems, $P\overline{V}$ is negligible, and we may set $\mu(T,P)$ equal to μ^0 (T,P = 0), without making serious errors.

Note: Because $P\overline{V}$ is small or negligible in comparison with RT, there is little difference between E and H and between A and G.

8.5.2 Entropy

The entropy is obtained at once from μ (T,P)

$$\overline{S} = -(\partial\mu/\partial T)_P$$
$$\approx \overline{S}^0(T, P = 0) - \alpha PV(T, 0)(1 - \tfrac{1}{2}\kappa P) \qquad (8\text{-}49a)$$
$$= \overline{S}^0(T, P = 0) \qquad (8\text{-}49b)$$

8.5.3 Enthalpy

$$\overline{H} = \mu + T\overline{S}$$
$$= \overline{H}^0(T, P = 0) + P\overline{V}(T, 0)(1 - \alpha T)(1 - \tfrac{1}{2}\kappa P) \qquad (8\text{-}50a)$$
$$\approx \overline{H}^0(T, P = 0) \qquad (8\text{-}50b)$$

CHAPTER 9

PHASE AND CHEMICAL EQUILIBRIA

In this chapter, we consider two types of equilibriums, equilibrium between heterogeneous phases and chemical reaction equilibrium. The treatment will not be exhaustive but will rather focus on procedures and topics not likely to be covered in elementary physical chemistry courses.

9.1 THE PHASE RULE

The phase rule reads

$$v = c + 2 - p \qquad (9\text{-}1)$$

where c stands for the number of components, p the number of phases, and v the variance (or degrees of freedom). The variance is defined as the number of *independent intensive* variables that can be specified in a heterogeneous equilibrium.

The phase rule is easy to derive. Let r denote the number components present and α the number of phases. In a particular phase, the following intensive variables are present: $T^{(\alpha)}, P^{(\alpha)}, x_1^{(\alpha)}, \ldots, x_{r-1}^{(\alpha)}$. Thus, there

---

Thermodynamics and Introductory Statistical Mechanics, by Bruno Linder
ISBN 0-471-47459-2 © 2004 John Wiley & Sons, Inc.

TABLE 9.1   Intensive Variables in α Phases

$T^{(1)} = T^{(2)} \ldots = T^{(\alpha)}$	$(\alpha - 1)$ equal signs
$P^{(1)} = P^{(2)} \ldots = P^{(\alpha)}$	"
$\mu_1^{(1)} = \mu_1^{(2)} \ldots = \mu_1^{(\alpha)}$	"
$\mu_r^{(1)} = \mu_r^{(2)} \ldots = \mu_r^{(\alpha)}$	"

are $(r + 1)$ intensive variables present in one phase. In all α phases, there are $\alpha(r + 1)$ intensive variables. However, not all variables are independent. For "normal" systems, which we are considering here, all phases have the same temperature, the same pressure, and the same chemical potential of each species, *i*. In short, numerous variables are equal to each other as shown in Table 9.1 Each time the values of two variables are set equal, the number of independent variables decreases by one. Table 9.1 shows that there are $(r + 2)$ lines, each line contributing $(\alpha - 1)$ equal signs. Thus the total number of equal signs is $(r + 2)(\alpha - 1)$, reducing the number of intensive variables by that amount. Accordingly, the number of *independent* intensive variables, i.e., the variance, is $v = -\alpha + r + 2$ or, in more standard notation (replacing α by p, and r by c), is the expression given by *Eq. 9-1*.

Note: We emphasize that this well-known formula for the phase rule *(Eq. 9-1)* holds only if the external force is a uniform pressure, the interphase surfaces are fully heat-conducting, deformable, and permeable, and there are no chemical reactions between the components.

Example
In a one-component system

$$
\begin{aligned}
v &= 2 \quad \text{if} \quad p = 1; \\
v &= 1 \quad \text{if} \quad p = 2; \text{ and} \\
v &= 0 \quad \text{if} \quad p = 3
\end{aligned}
\qquad (9\text{-}2)
$$

Figure 9.1 shows a typical phase diagram of a one-component system. (Phase diagrams of multi-component systems will not be covered in this book. Readers who want to study these are advised to consult some good undergraduate physical chemistry textbooks.)

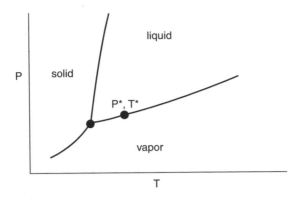

Figure 9.1   Phase diagram of a one-component system.

9.2 THE CLAPEYRON EQUATION

Consider a one-component, two-phase system in equilibrium. Suppose the system is characterized by the point P^*, T^* on one of the coexisting curves (see Figure 9.1). The change to a neighboring point $P^* + dP$, $T^* + dT$ will necessitate also changing the chemical potential from μ to $\mu + d\mu$. We know from elementary considerations (or from Legendre transformations) that for a single phase, α,

$$d\mu^\alpha = -(S^\alpha/n)dT + (V^\alpha/n)dP = -\overline{S}^\alpha dT + \overline{V}^\alpha dP \qquad (9\text{-}3)$$

If two phases, α and β coexist, then $\mu^\alpha = \mu^\beta$ and $\mu^\alpha + d\mu^\alpha = \mu^\beta + d\mu^\beta$. Thus,

$$d\mu^\alpha = d\mu^\beta$$
$$-\overline{S}^\alpha dT + \overline{V}^\alpha dP = -\overline{S}^\beta dT + \overline{V}^\beta dP \qquad (9\text{-}4)$$

or

$$(dP/dT)_{\text{coex}} = (\overline{S}^\beta - \overline{S}^\alpha)/(\overline{V}^\beta - \overline{V}^\alpha)$$
$$= \Delta\overline{S}/\Delta\overline{V} \qquad (9\text{-}5)$$

This is the Clapeyron Equation. A more useful form is

$$(dP/dT)_{\text{coex}} = \Delta\overline{H}/(T\Delta\overline{V}) \qquad (9\text{-}6)$$

9.3   THE CLAUSIUS-CLAPEYRON EQUATION

The Clausius-Clapeyron Equation is an approximation, used frequently in discussing solid-vapor and liquid-vapor equilibria. The basic idea is that $\overline{V}_{solid}$ and $\overline{V}_{liquid}$ are small in comparison with the volume of the gas, and so can be neglected. For liquid-vapor, we have

$$dP/dT = (\overline{H}_v - \overline{H}_l)/T(\overline{V}_v - \overline{V}_l) \approx \Delta\overline{H}_{vap}/T\overline{V}_v \qquad (9\text{-}7)$$

Treating the vapor as an ideal gas, we get

$$dP/dT \approx \Delta\overline{H}_{vap}P/TRT \qquad (9\text{-}8a)$$

and assuming that $\Delta\overline{H}_{vap}$ varies negligibly with temperature, we obtain

$$\int dP/P \approx \int (\Delta\overline{H}_{vap}/RT^2)dT \approx \frac{\Delta\overline{H}_{vap}}{R}\int_{T_1}^{T_2}\frac{dT}{T^2} \qquad (9\text{-}8b)$$

which gives, the Clausius-Clapeyron equation

$$\ln(P_2/P_1) \approx \Delta\overline{H}_{vap}/R(1/T_1 - 1/T_2) \approx \Delta\overline{H}_{vap}/R[(T_2 - T_1)/T_1T_2] \quad (9\text{-}9)$$

Similarly, for solid-vapor equilibrium,

$$\ln(P_2/P_1) \approx \Delta\overline{H}_{sub}/R[(T_2 - T_1)/T_1T_2] \qquad (9\text{-}10)$$

The Clapeyron and Claussius-Clapeyron equations are powerful tools for constructing phase diagrams of the types shown in Figure 9.1. The behavior of the coexistence curves can actually be inferred from qualitative considerations of the relative magnitudes of the $\Delta\overline{H}$ and $\Delta\overline{V}$ values of the coexistence curves. For solid-vapor equilibrium and for liquid-vapor equilibrium, the $\Delta\overline{V}$ values are essentially of the same order of magnitude, but the enthalpy of sublimation always exceeds the enthalpy of vaporization, that is, $\Delta\overline{H}_{subl} > \Delta\overline{H}_{vap}$, and so the slope of the solid-vapor curve is expected to be greater than that of the liquid-vapor curve. The solid-liquid $\Delta\overline{H}$, that is, $\Delta\overline{H}_{fusion}$, is generally small, but the $\Delta\overline{V}_{fusion}$ is usually very small (and sometimes negative). The solid-liquid slope is likely to be very large. Thus, one can expect the slope of the solid-vapor curve to be larger than that of the liquid-vapor curve and the slope of the solid-liquid curve to be largest (and sometimes negative). These predictions are borne out by experiment.

9.4 THE GENERALIZED CLAPEYRON EQUATION

The foregoing analysis of the Clapeyron Equation is based on the assumption that the system is "normal" and thus there is uniform temperature, uniform pressure, and uniform chemical potential throughout the phases. What if there are restrictions, for example, when the inter-phase boundary is semi-deformable so that the pressure on the liquid (the applied pressure!), P^l, is not equal to the vapor pressure, P^v? How can the Clapeyron Equation be modified to apply to this situation? This is, in principle, possible by subjecting the liquid to a membrane at extremely high pressures; in practice, however, it can be accomplished by adding to the vapor phase an inert gas, which is insoluble in the liquid phase. Under such circumstances, the vapor consists of two (or more) species; the liquid phase has only one substance, say, species i. Yet, the chemical potentials of i, that is, μ_i^v and μ_i^l, are the same in both phases, as are also the temperatures but not the pressures: $P_i^v \neq P_i^l$. Dropping the subscript i to simplify the notation, we get

$$-\overline{S}^l dT + \overline{V}^l dP^l = -\overline{S}^v dT + \overline{V}^v dP^v \qquad (9\text{-}11)$$

which yields

1) $(\partial P^v/\partial T)_{P^l} = \Delta \overline{H}_{vap}/T\overline{V}^v = [(\overline{S}^v - \overline{S}^l)/\overline{V}^v]$ $(9\text{-}12)$

2) $(\partial P^v/\partial P^l)_T = \overline{V}^l/\overline{V}^v$ $(9\text{-}13)$

3) $(\partial P^l/\partial T)_{P^v} = -\Delta \overline{H}_{vap}/T\overline{V}^l = [-(\overline{S}^v - \overline{S}^l)/\overline{V}^l]$ $(9\text{-}14)$

EXAMPLE

Under what applied pressure would water at 99.999°C or 373.149 K exhibit a vapor pressure of 1 atm? The specific heat of vaporization of water is 540 cal; the specific volume is 1 cm^3.

SOLUTION

We know that under normal conditions the applied or liquid pressure is the same as the vapor pressure, namely 1 atm, at the normal boiling point of 373.150 K. Holding P^v constant and integrating Eq. 9.14) from T = 373.150 K to 373.149 K yields the final applied pressure, P^l

$$P^l - 1 \approx -(\Delta \overline{H}_{vap}/\overline{V}^l) \ln(373.149/373.150) \qquad (9\text{-}15)$$

9.5 CHEMICAL EQUILIBRIUM

Thermodynamic aspects of chemical reactions are almost always formulated on the basis of the Gibbs free energy. This is not absolutely necessary. We can also arrive at the fundamental relations between the equilibrium constant and the thermodynamic properties of the reacting substances from other thermodynamic potentials. Because our focus in deriving equilibrium and stability conditions in the preceding chapters has been on the internal energy E, we shall show here that E can be used to derive equilibrium conditions for chemical reactions.

Consider the reaction,

$$aA + bB + \cdots = cC + dD + \cdots \qquad (9\text{-}16)$$

written simply as

$$\Sigma_i \nu_i d\chi_i = 0 \qquad (9\text{-}17)$$

which is interpreted to mean that the stochiometric coefficient ν_i is positive (+) when the substance χ_i is a product and negative (−) when the substance χ_i is a reactant.

Introducing the variable ξ, called *the progress variable* or *degree of advancement*, the change in mole numbers becomes $dn_i = \nu_i d\xi$, meaning that, for example, when $dn_A = \nu_A d\xi$, dn_A moles of A disappear and when $dn_C = \nu_C d\xi$, dn_C moles of C are created. Using the energy criterion for equilibrium discussed earlier, we require that

$$\delta^{(1)} E_{S,V} \geq 0 \qquad (9\text{-}18)$$

Note that in this expression the constraint that the n_i is constant is missing. This is so because here we are dealing with chemical reactions and the mole numbers are changing. Because

$$\delta E = T\delta S - P\delta V + \Sigma_i \nu_i \mu_i \delta n_i \qquad (9\text{-}19)$$

we have

$$(\delta^{(1)}E)_{S,V} = \Sigma_i(\partial E/\partial n_i)_{S,V,n_{j\neq i}} dn_i = \Sigma_i \mu_i \delta n_i \geq 0 \qquad (9\text{-}20)$$

$$= \Sigma_i \mu_i \nu_i d\xi \geq 0 \qquad (9\text{-}21)$$

Note: The quantity $\Sigma_i \nu_i \mu_i$, denoted as $\Delta\mu$, is sometimes called "reaction potential," and its negative, $-\Delta\mu = -\Sigma_i \nu_i \mu_i$, is called "affinity."

Now, let us consider a virtual variation $d\xi$, which takes the reacting mixture away from equilibrium under the constraints of constant S and V. If $d\xi > 0$, then $\Sigma_i \nu_i \mu_i \geq 0$, indicating that at equilibrium the reaction potential was either zero or positive. If $d\xi < 0$, then $\Sigma_i \nu_i \mu_i \leq 0$, indicating that the reaction potential at equilibrium was zero or negative. Obviously, the only consistent result is

$$\Delta\mu = \Sigma_i \nu_i \mu_i = 0 \qquad (9\text{-}22)$$

The same identical result would have been obtained had we used any of the other thermodynamic potentials: $H(S, P; n_i)$, $A(T, V; n_i)$, or $G(T, P; n_i)$. In short, the results are independent of the various constraints, and *Eq. 9-22* is the appropriate criterion for chemical equilibrium.

9.6   THE EQUILIBRIUM CONSTANT

The foregoing thermodynamic treatment of chemical equilibrium must, from a chemist's point of view, be disappointing because there is no mention of an equilibrium constant. Before we can define an equilibrium constant, we must introduce a new concept, namely, the activity a_i. This concept will be discussed in more detail in Chapter 10. For the time being, let us think of the activity as an "effective" concentration or "effective" pressure. (For gases, the activity is the same as the fugacity.)

In general, the chemical potential of a given species is a function of T and P and the mole fractions of all species. The activity is defined by writing

$$\mu_i(T, P; x_1, x_2, \ldots, x_{r-1}) = \mu_i^0(T, P) + RT \ln a_i(T, P; x_1, \ldots, x_{r-1}) \quad (9\text{-}23)$$

(For a gas, the standard chemical potential, μ_i^0, is independent of pressure.)

It is obvious that the activity will depend on the choice of the standard state. The choice is strictly a matter of convention, which varies from gases to pure substances and from pure substances to solutions, etc.

Substituting *Eq. 9-23* into *Eq. 9-22* gives

$$\Delta\mu = \Delta\mu^0 + RT\ \Sigma_i v_i \ln a_i = 0 \qquad (9\text{-}24)$$

Consequently,

$$\Delta\mu^0 = \Sigma_i v_i \mu_i^0 = -RT\ \ln\ \Pi_i a_i^{v_i} \equiv -RT \ln K_a \qquad (9\text{-}25a)$$

where Π_i stands for the product $a_1^{v_1} a_2^{v_2} \cdots$

Accordingly

$$K_a = e^{-\Delta\mu^0/RT} \qquad (9\text{-}25b)$$

where K_a is the equilibrium constant defined here in terms of the activities.

It should be noted that the activities are dimensionless. Also, it is important to keep in mind that the equilibrium constant is not related to the reaction potential of the real mixture (which is zero) but to the reaction potential of the substances in their standard states.

Finally, because $\Delta\mu^0$ is a free energy change, the Gibbs-Helmholtz Equation applies

$$(\partial(\Delta\mu^0/T)/\partial T)_P = -\Delta\overline{H}^0/RT^2 \qquad (9\text{-}26a)$$

where

$$\Delta\overline{H}^0 = \Sigma_i v_i \overline{H}_i^0 \qquad (9\text{-}26b)$$

Using *9-25a* and *9-26b* yields

$$(\partial \ln K_a/\partial T)_P = \Delta\overline{H}^0/RT^2 \qquad (9\text{-}27)$$

It also follows from $(\partial\Delta\mu^0/\partial P)_T = \Delta\overline{V} = \Sigma_i v_i \overline{V}_i^0$ that

$$(\partial \ln K_a/\partial P)_T = -\Delta\overline{V}^0/RT \qquad (9\text{-}28)$$

CHAPTER 10

SOLUTIONS—NONELECTROLYTES

We have already discussed mixtures of *gases*. In Chapter 10, we present a more general discussion of mixtures and treat in some detail binary solutions of *solids in liquids*, with special emphasis on *dilute* solutions. Only nonelectrolytes will be considered. The main focus in this chapter, as in the previous, will be on the generality of the method.

10.1 ACTIVITIES AND STANDARD STATE CONVENTIONS

We have introduced the concept of activity in *Equation 9-23*. This concept is really a figment of the mind. The only quantity that has a thermodynamic base is the chemical potential of the system, μ_i. The standard chemical potential, μ_i^0, and activity, a_i, are arbitrary and depend on the convention adopted for defining them. In mixtures, it is customary to choose a reference or standard state that is independent of mole fractions but dependent on temperature and often also on pressure.

10.1.1 Gases

Recall that for gases

$$\mu_i(T, P) = \mu_i^0(T) + RT \ln f_i/P^0 \qquad (10\text{-}1)$$

Thermodynamics and Introductory Statistical Mechanics, by Bruno Linder
ISBN 0-471-47459-2 © 2004 John Wiley & Sons, Inc.

we take the standard state (as before) to be

$$\mu_i^0(T, P) = \mu_i^0(T) \qquad (10\text{-}2)$$

which then defines the activity

$$a_i = f_i / P^0 \qquad (10\text{-}3)$$

10.1.2 Pure Liquids and Solids

Here, we take the standard state to be the state of the pure substance i. So,

$$\mu_i^0(T, P) = \mu_i^\bullet(T, P) \qquad (10\text{-}4a)$$

and

$$a_i = 1 \qquad (10\text{-}4b)$$

10.1.3 Mixtures

There are three ways to define standard states: (1) in terms of mole fractions, x_i (rational basis); (2) in terms of molality, m_i [i.e., moles of solutes in 1 kg of solvent (molal basis)]; and (3) in terms of molarity, c_i [i.e., moles of solutes in 1 liter of solution (molar basis)]. The activity is often written as a product of an activity coefficient and a concentration factor: $a_i = \gamma_i^{(x)} x_i$ (rational basis), $a_i = \gamma_i m_i^{(m)}$ (molal basis), and $a_i = \gamma_i^{(c)} c_i$ (molar basis). We will use only the rational basis (x) here to illustrate the treatment of solutions.

Let species "1" refer to the solvent and the other species, $i = 2$, $i = 3$, etc., refer to the solutes. In some solutions, no fundamental difference exists between solute and solvent: the species in *greater abundance* is called the *solvent*. This is generally true of liquid-liquid mixtures. In other cases, such as solid-liquid mixtures there is a discernable difference between solute and solvent. These two types of solutions behave so differently that they are generally treated by different conventions. By one convention (denoted here as Con I), the standard chemical potential, μ^0, is the chemical potential of the pure substance, regardless of whether the substance is the *solute* or the *solvent*. By the other convention (denoted here as Con II), the standard chemical potential of the solvent is that of the pure solvent, but the standard chemical potential of the solute is not that of the *pure* solute; rather, it is the chemical potential of a state, which is not a true state of the solute.

10.1.3.1 Liquid–Liquid Solutions—Convention I (Con I)
Here, we take the standard state to be the state of the pure substance, regardless whether the substance is the "solvent" or a "solute". Thus, for each species, i

$$\mu_i^0(T, P) = \mu_i^\bullet(T, P) \qquad (10\text{-}5)$$

$$a_i = \gamma_i x_i \qquad (10\text{-}6)$$

10.1.3.2 Solid-Liquid Solutions—Convention II (Con II)
Here, the chemical potential of the pure solvent (species "1") is taken to be the standard chemical potential. Thus

$$\mu_1^0(T, P) = \mu_1^\bullet(T, P) \qquad (10\text{-}7)$$

$$\mu_i^0(T, P) \neq \mu_i^\bullet(T, P) \quad i \neq 1$$

$$= \lim_{x_i \to 0} (\mu_i - RT \ln x_i) \qquad (10\text{-}8)$$

Note: This Con II is used almost exclusively for ideally dilute solutions (defined below). The solute mole fraction for such solutions approaches zero, and the solvent mole fraction approaches one. The activity coefficient of the solvent is essentially one, and the activity of the solvent is basically the same as the mole fraction, i.e., $a_1 \approx x_1$. The reason solutes and solvents behave so different in infinitely dilute solutions is because the environment of a solvent molecule is basically the same as that for the pure solvent, but the environment of the solute molecule differs radically from that of the pure solute.

10.2 IDEAL AND IDEALLY DILUTE SOLUTIONS; RAOULT'S AND HENRY'S LAWS

10.2.1 Ideal Solutions

In elementary treatments, it is standard to define an ideal solution as one in which each component obeys Raoult's Law. Raoult's Law states that the fugacity of each component i (regardless of whether it is a solute or the solvent) is equal to the mole fraction of i in the liquid phase times the fugacity of pure i, i.e.

$$f_i = x_i f_i^\bullet \qquad (10\text{-}9)$$

or, (as is more common although less precise), in terms of the partial pressure of i, P_i, and pressure of the pure substance, $P_i^\bullet$

$$P_i = x_i \, P_i^\bullet \qquad (10\text{-}10)$$

Ideal solutions can be defined in another way, by focusing on the properties of the solution rather than on the vapor. From this standpoint, an ideal solution is defined as one whose components satisfy the relation

$$\mu_i(T, P, x_i) = \mu_i^\bullet(T, P) + RT \ln x_i \quad i = 1, 2, \dots \qquad (10\text{-}11)$$

We assume that the relation holds over the entire range of compositions. To be specific, let us consider a liquid-vapor equilibrium. We must have

$$\mu_i^v = \mu_i^l \qquad (10\text{-}12)$$

$$\mu_i^{0v}(T) + RT \ln (f_i/P^0) = \mu_i^{0l}(T, P) + RT \ln x_i \qquad (10\text{-}13)$$

Thus

$$f_i/P^0 = x_i \exp[(\mu_i^{0l} - \mu_i^{0v})/RT] \qquad (10\text{-}14a)$$

$$= x_i \exp[(\mu_i^{\bullet l} - \mu_i^{0v})/RT] \qquad (10\text{-}14b)$$

For pure i

$$f_i^\bullet/P^0 = \exp[(\mu_i^{\bullet l} - \mu_i^{0v})/RT] = \exp[(\mu_i^{\bullet l} - \mu_i^{\bullet v})/RT] \qquad (10\text{-}15)$$

This shows that

$$f_i = x_i f_i^\bullet \qquad (10\text{-}16)$$

which is Raoult's Law.

10.2.2 Ideally Dilute Solutions

For all solutions, when sufficiently dilute, the solvent obeys Raoult's Law but the solute does generally not. Rather, the solute obeys *Henry's Law*:

$$f_i = x_i \, k_H \quad i \neq 1 \qquad (10\text{-}17)$$

where k_H is Henry's Law constant. In general, k_H differs from $f_i^\bullet$. When they are equal, the solution is ideal.

It can be shown that, in the range in which the solvent obeys Raoult's Law, the solute will obey Henry's Law. In elementary treatments, this is normally proved by applying the fundamental relation $\Sigma_i\, n_i d\mu_i = 0$ (the Gibbs-Duhem equation) and differentiating the chemical potentials with respect to x_i. Here, we establish this result in another way. We define an *ideally dilute solution* as one in which each component (in the liquid phase) obeys the relation

$$\mu_i = \mu_i^0 + RT \ln x_i \qquad (10\text{-}18)$$

Note: This expression is of the same form as the expression for an ideal solution, but there is a difference In an *ideal solution*, $\mu_i^0 = \mu_i^{\bullet}$, irrespective of whether i represents the solute or the solvent, but, in an *ideally dilute solution*, $\mu_i^0 \neq \mu_i^{\bullet}$ when i is not the solvent.

Let us assume again that we are dealing with a binary, liquid-vapor, system. We equate the chemical potential of each of the species in the liquid with that in the vapor and obtain

$$\mu_1^{\bullet l} + RT \ln x_1 = \mu_1^{0v}(T) + RT \ln(f_1/P^0) \qquad (10\text{-}19)$$

$$\mu_2^{0l} + RT \ln x_2 = \mu_2^{0v}(T) + RT \ln(f_2/P^0) \qquad (10\text{-}20)$$

Accordingly

$$\ln(f_2/P^0) = (\mu_2^{0l} - \mu_2^{0v})/RT + \ln x_2 \qquad (10\text{-}21)$$

or

$$f_2/P^0 = x_2 \exp[(\mu_2^{0l} - \mu_2^{0v})/RT] \qquad (10\text{-}22)$$

By identifying the exponential factor on the right with k_H/P^0

$$k_H/P^0 \equiv \exp[(\mu_2^{0l} - \mu_2^{0v})/RT] \qquad (10\text{-}23)$$

we get

$$f_2 = x_2 k_H \qquad (10\text{-}24)$$

which is Henry's Law.

Note: k_H is independent of x_2 but is dependent on P because μ_2^{0l} depends on P (as well as on T). But the dependence on P is weak and varies slowly so that k_H is practically constant except at very high pressures.

Note: $k_H/P^0 \neq f_2^{\bullet}/P^0$, the latter is equal to $\exp\left[(\mu_2^{\bullet l} - \mu_2^{0v})/RT\right]$

Thus

$$f_1^0 = f_1^{\bullet}$$

but

$$f_2^0 \neq f_2^{\bullet} \tag{10-25}$$

10.3 THERMODYNAMIC FUNCTIONS OF MIXING

In general

$$\mu_i(T, P; x_i) = \mu_i^0(T, P) + RT \ln \gamma_i x_i \tag{10-26}$$

$$\Delta G_{mix} = G - \Sigma_i n_i G_i^{\bullet} = \Sigma_i n_i^{\bullet}(\mu_i - \mu_i^{\bullet}) \tag{10-27}$$

10.3.1 For Ideal Solutions

$$\Delta G_{mix} = RT \, \Sigma_i n_i \ln x_i \tag{10-28}$$

$$\Delta S_{mix} = -R \, \Sigma_i n_i \ln x_i \tag{10-29}$$

$$\Delta H_{mix} = 0 \tag{10-30}$$

$$\Delta V_{mix} = 0 \tag{10-31}$$

10.3.2 For Nonideal Solutions

For *Liquid-liquid mixtures* (Con I)

$$\mu_i^0 = \mu_i^{\bullet} \text{ all } i \tag{10-32}$$

$$\Delta G_{mix} = \Sigma_i n_i(\mu_i - \mu_i^0) + \Sigma_i n_i(\mu_i^0 - \mu_i^{\bullet}) \tag{10-33a}$$

$$\Delta G_{mix} = RT \, \Sigma_i n_i \ln \gamma_i x_i + 0 \tag{10-33b}$$

$$= RT \, \Sigma_i n_i \ln x_i + RT \, \Sigma_i n_i \ln \gamma_i(T, P; x_1, \ldots, x_r) \tag{10-34a}$$

$$\equiv \Delta G_{mix}^{ideal} + \Delta G_{mix}^{excess} \tag{10-34b}$$

$$\Delta S_{mix} = -R\Sigma_i n_i \ln x_i - [\partial(\Delta G_{mix}^{excess}/\partial T)]_{P,n} \tag{10-35}$$

$$= \Delta S_{mix}^{ideal} + \Delta S_{mix}^{excess}$$

$$\Delta H_{mix} = 0 + [\partial(\Delta G_{mix}^{excess}/T)\partial(1/T)]_{P,n} \tag{10-36}$$

$$\Delta V_{mix} = 0 + [\partial \Delta G_{mix}^{excess}/\partial P]_{T,n} \tag{10-37}$$

For *Solid-Liquid mixtures* (Con II)

$$\mu_1^0 = \mu_1^\bullet \text{ solvent} \tag{10-38a}$$

$$\mu_i^0 \neq \mu_i^\bullet \text{ solute } i = 2, 3, \ldots \tag{10-38b}$$

$$\Delta G_{mix} = \Sigma_i n_i (\mu_i - \mu_i^0) + \Sigma_i n_i (\mu_i^0 - \mu_i^\bullet) \tag{10-39a}$$

$$= RT \Sigma_i n_i \ln x_i + RT \Sigma_i n_i \ln \gamma_i + \Sigma_{i \neq 1} n_i (\mu_i^0 - \mu_i^\bullet) \tag{10-39b}$$

$$\Delta S_{mix} = -R \Sigma_i n_i \ln x_i - R \Sigma_i n_i \ln \gamma_i - RT \Sigma_i n_i (\partial \ln \gamma_i / \partial T)_{P,n}$$

$$- \Sigma_{i \neq 1} n_i (\overline{S}_i^0 - \overline{S}_i^\bullet) \tag{10-40}$$

$$\Delta H_{mix} = RT \Sigma_i n_i (\partial \ln \gamma_i / \partial T)_{P,n} + \Sigma_{i \neq 1} n_i (\overline{H}_i^0 - \overline{H}_i^\bullet) \tag{10-41}$$

$$\Delta V_{mix} = RT \Sigma_i n_i (\partial \ln \gamma_i / \partial P)_{T,n} + \Sigma_i n_{i \neq 1} (\overline{V}_i^0 - \overline{V}_i^\bullet) \tag{10-42}$$

10.4 COLLIGATIVE PROPERTIES

The lowering of the chemical potential of the solvent in solutions has a profound effect on several properties of the system, collectively referred to as *colligative properties.* They include

1) lowering of solvent vapor pressure
2) depression of freezing point
3) elevation of boiling point, and
4) osmotic pressure.

The word colligative means that the properties depend only on the quantity (amount) and not on the identity of the solutes. Colligative properties are most commonly applied to very dilute (ideally dilute) solutions and usually to solutions in which the solute (we will here focus only on one) is nonvolatile. We will refer to the solvent as A and the solute as B. Nonionic substances only will be treated.

10.4.1 Lowering of Solvent Vapor Pressure

This topic was discussed already in Section 10.2 in some detail and need not be repeated here. In all solutions, $\mu_A = \mu_A^0 + RT \ln a_A$ and the nonideality of the vapor can be ignored; thus we can set $a_A = P_A / P^0$. Employing Raoult's Law, $P_A = x_A P^\bullet$ gives $a_A = x_A P^\bullet / P^0$ and for the pure solvent $a_A^\bullet = P_A^\bullet / P_A^0$. Because $\mu^0 = \mu^\bullet$, we conclude that

$$a_A < a_A^\bullet, \quad \mu_A < \mu_A^\bullet, \quad \text{and} \quad P_A < P_A^\bullet \tag{10-43}$$

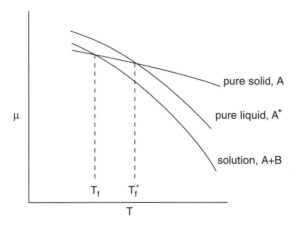

Figure 10.1   Plot of μ vs. T for pure solid, pure liquid, and solution solvent.

10.4.2 Freezing Point Depression

Figure 10.1 depicts the variation of μ with T for a pure solid A, pure liquid A, and solution, A and B. The solution is very dilute; the solute B is non-volatile. The normal freezing point, that is, the freezing point of the pure liquid, is denoted as $T_f{}^*$. The freezing point of the solution is denoted as T_f. It is apparent from Figure 10.1 that T_f is lower than T_f^*.

For the pure solid

$$\mu_A(s) = \mu_A^0(s) = \mu_A^\bullet(s) \qquad (10\text{-}44)$$

For the solvent

$$\mu_A(l) = \mu_A^\bullet(l) + RT \ln x_A \qquad (10\text{-}45)$$

and for solid-liquid equilibrium

$$\mu_A(s) = \mu_A(l) \qquad (10\text{-}46)$$
$$\mu_A^\bullet(s) = \mu_A^\bullet(l) + RT \ln x_A$$
$$= \mu_A^\bullet(l) + RT \ln(1 - x_B)$$
$$\approx \mu_A^\bullet(l) - RT x_B \qquad (10\text{-}47)$$

Let us define $\Delta\mu = \mu_A^\bullet(l) - \mu_A^\bullet(s)$. It is clear that, when $T = T_f^*$, $x_B = 0$ and so

$$\Delta\mu(T_f^*) = 0 \qquad (10\text{-}48)$$

On the other hand, when $T = T_f$

$$\Delta\mu(T_f) = RTx_B \qquad (10\text{-}49)$$

Applying the Gibbs-Helmholtz equation yields

$$[\partial(\Delta\mu/T)/\partial T]_P = -\Delta\overline{H}_{fus}/T^2 \qquad (10\text{-}50)$$

where $\Delta\overline{H}_{fus}$ is the molar enthalpy of fusion.

Assuming that $\Delta\overline{H}_{fus}$ is essentially constant in the small temperature range between T_f^* and T_f, integration between the limits T_f and T_f^* yields

$$\Delta\mu(T_f^*) - \Delta\mu(T_f) = -\Delta\overline{H}_{fus}\ln(T_f^*/T_f) \qquad (10\text{-}51)$$

Using Eqs. 10.48 and 10.49 and defining $\Delta T_f = T_f^* - T_f$, we obtain

$$-RT_f x_B = -\Delta\overline{H}_{fus}\ln(1 + \Delta T_f/T_f) \qquad (10\text{-}52)$$

which, upon expanding

$$\ln(1 + \Delta T_f/T_f) \approx \Delta T_f/T_f \qquad (10\text{-}53)$$

gives

$$x_B \approx (\Delta\overline{H}_{fus}/R)(\Delta T_f/T_f^{*2}) \qquad (10\text{-}54)$$

or

$$\Delta T_f = x_B[RT_f^{*2}/\Delta\overline{H}_{fus}] \qquad (10\text{-}55)$$

The quantity within square brackets in *Equation 10-55* is freezing point constant, which was obtained here on the rational basis. It is more common to use the molal basis. This can readily be obtained by replacing x_B by $n_B/(n_A + n_B) \approx n_B/n_A$ since n_B is much smaller than n_A. By taking the weight of the solvent to be 1 kg and expressing the molar weight (M_A) in kg/mol, we have $n_A = 1/M_A$ and the molality of B is $m_B = n_B M_A$. Accordingly

$$\Delta T_f = [RT_f^{*2}M_A/\Delta\overline{H}_{fus}]m_B = K_f m_B \qquad (10\text{-}56)$$

For water, the cryogenic constant, K_f, is

$$\begin{aligned} K_f &= (18.015 \times 10^{-3}\,\text{kg/mol})(8.314\,\text{J}\cdot\text{K}^{-1}\cdot\text{mol}^{-1}) \\ &\quad \times (273.15\,\text{K})^2/(6008\,\text{J/mol}) \\ &= 1.86\,\text{K}\cdot\text{kg}\cdot\text{mol}^{-1} \qquad (10\text{-}57) \end{aligned}$$

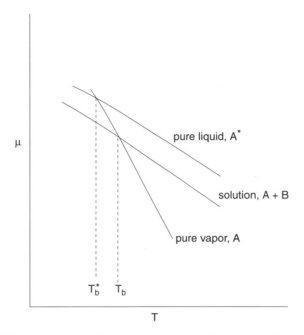

Figure 10.2   Plot of μ vs. T for pure, vapor pure liquid and solution solvent.

10.4.3   Boiling Point Elevation

Figure 10.2 is a schematic diagram of the variation of μ with T of the pure vapor, the pure solvent and the solvent in solution. There are three differences between this diagram and the depression of freezing point diagram (Fig. 10.1). The pure solid curve has been replaced by the pure vapor curve, and the normal boiling point, T_b^*, lies below the boiling point, T_b, of the solution. It is obvious that the transition between solid and liquid has to be replaced by the transition from vapor to liquid, or by $-\Delta H_{vap}$, which has the opposite sign to ΔH_{fus}. If we define $\Delta T_b = T_b - T_b^*$, which also has the opposite sign to ΔT_f, it is clear that the expression for the ebullioscopy constant ought to be the same as for the cryogenic constant, except for the replacement of $\Delta \overline{H}_{fus}$ by $\Delta \overline{H}_{vap}$ and ΔT_f by ΔT_b. The final expression for the elevation of the boiling point becomes

$$\Delta T_b = [RT_b^{*2} M_A / \Delta \overline{H}_{vap}]\, m_B = K_b\, m_B \qquad (10\text{-}58)$$

EXERCISE

Derive *Eq. 10-58* and evaluate the ebullioscopy constant K_b for H_2O.

$$(\Delta H_{vap} = 40.656\,kJ/mol \text{ and } T_b^* = 373.15\,K)$$

10.4.4 Osmotic Pressure

In discussing the depression of the freezing point and elevation of the boiling point, it was tacitly assumed that the pressure was constant. The treatment of osmotic pressure is based on the notion that the temperature is constant, but the properties of the solution change with pressure.

As a working model, consider two vessels separated by a semi-permeable membrane. On the left (Fig. 10.3) is pure A. On the right is a solution of solvent A and solute B. The membrane is permeable to A but not to B. Obviously, the chemical potential of A on the right is lower than the chemical potential on the left. Matter (A) will flow from left to right, causing the liquid on the right to rise. *The extra pressure, Π, that must be applied to prevent liquid flow from left to right is the osmotic pressure.* If P denotes the atmospheric pressure, then at equilibrium

$$\mu_{A,L}(P) = \mu_{A,R}(P + \Pi) \qquad (10\text{-}59)$$

Again, assuming that we are dealing with a very dilute solution, we can write

$$\mu_{A,L}(P) = \mu_{A,L}^{\bullet}(P) \qquad (10\text{-}60)$$

$$\mu_{A,R}(P + \Pi) = \mu_{A,R}^{\bullet}(P + \Pi) + RT \ln x_A \qquad (10\text{-}61)$$

and thus,

$$\mu_{A,L}^{\bullet}(P) = \mu_{A,R}^{\bullet}(P + \Pi) + RT \ln x_A \qquad (10\text{-}62)$$

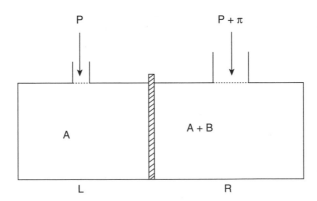

Figure 10.3 Schematic representation of two vessels, separated by a semi-permeable membrane. P is the atmospheric pressure, and Π is the extra (osmotic) pressure necessary to prevent flow of solvent a from L to R.

Recall that

$$d\mu = -\overline{S}dT + \overline{V}dP \text{ and thus at constant T } (\partial\mu/\partial P)_T$$
$$= \overline{V} \text{ (where } \overline{V} \text{ refers to the molar volume). Hence}$$

$$d\mu^\bullet_{A,R} = \overline{V}^\bullet_A dP \quad \text{(constant T)} \tag{10-63}$$

Integrating between the limits P and $(P + \Pi)$ and assuming that the liquid is incompressible, we get

$$\mu^\bullet_{A,R}(P + \Pi) - \mu^\bullet_{A,R}(P) = \overline{V}^\bullet_A \int_P^{P+\Pi} dP' = \overline{V}^\bullet_A \Pi \tag{10-64}$$

Thus

$$\mu^\bullet_{A,L}(P) = \mu^\bullet_{A,R}(P) + \overline{V}^\bullet_A\Pi + RT \ln x_A \tag{10-65}$$

Because $\mu^\bullet_{A,L}(P) = \mu^\bullet_{A,R}(P)$, we have

$$\Pi = (-RT/\overline{V}^\bullet_A) \ln x_A = (-RT/\overline{V}_A) \ln(1 - x_B) \approx (RT/\overline{V}^\bullet_A) x_B \tag{10-66}$$

In very dilute solutions, $x_B = n_B/(n_A + n_B) \approx n_B/n_A$ and so

$$\Pi = (RT/\overline{V}^\bullet_A)(n_B/n_A) = RT \, n_B/V^\bullet_A \tag{10-67}$$

where $V^\bullet_A$ is the actual volume of the fluid; thus

$$\Pi = RT \, c_B \tag{10-68}$$

where c_B is the concentration of B in the right vessel. This equation is also known as the van 't Hoff equation.

Note: Osmotic pressure measurements are useful to obtain molecular weights of high polymers because their concentrations, c_B, are exceptionally low. For example, a solution of 1 g of solute B in 100 g of H_2O gives $\Delta T_f = -0.002°C$, but $\Pi = 19$ Torr at 25°C.

CHAPTER 11

PROCESSES INVOLVING WORK OTHER THAN PRESSURE-VOLUME WORK

It was mentioned earlier that work could take on various forms. Continuing to define work when *done* on the system to be *positive*, we may represent the element of work as

$$dw = \Sigma_i X_i dx_i \qquad (11\text{-}1)$$

where X_i is a generalized force and x_i a generalized displacement. We already gave a summary of various types of work in Chapter 2. Here, we will be concerned with three types of work:

1) $Xdx \leftrightarrow -PdV$ pressure-volume work; P = pressure and V = volume
2) $Xdx \leftrightarrow \sigma d\mathcal{A}$ film enlarging; σ = surface tension and $\mathcal{A}$ = area
3) $Xdx \leftrightarrow fdL$ stretching rubber; f = tensile strength and L = length

$$(11\text{-}2)$$

These forms can be used separately or in combination. Although, strictly speaking, work associated with surface enlargement or wire stretching cannot be treated independently of pressure-volume work, the effect is so small that it is generally an excellent approximation to ignore pressure-volume effect. For example, if a piece of rubber of length L and cross-section $\mathcal{A}$

Thermodynamics and Introductory Statistical Mechanics, by Bruno Linder
ISBN 0-471-47459-2 © 2004 John Wiley & Sons, Inc.

is stretched, then as L increases $\mathcal{A}$ will decrease, so that the volume change is small or negligible. Thus, in first approximation, the pressure-volume part of the work may be neglected. The same holds true for surface enlargement.

Still, it is instructive to consider more than one type of work because this clearly illustrates the generality of the approach. We will divide the treatment into two parts. In Section 11.1 we will discuss P-V work in combination with one other type. In Section 11.2, we will concentrate on one type of work and discuss in some detail the application of thermodynamices to P-V work, $\sigma\mathcal{A}$ work, and fL work.

11.1   P-V WORK AND ONE OTHER TYPE OF WORK

Using $E = E(S, V, y, n_i)$ gives

$$dE = TdS - PdV + Ydy + \Sigma_i\mu_i dn_i \qquad (11\text{-}3)$$

and

$$E = TS - PV + Yy + \Sigma_i\mu_i n_i \qquad (11\text{-}4)$$

Using Legendre transformations or standard definitions ($H = E + PV$, $A = E - TS$, $G = H - TS$), yields

$$dH = TdS + VdP + Ydy + \Sigma_i\mu_i dn_i \qquad (11\text{-}5)$$

$$dA = -SdT - PdV + Ydy + \Sigma_i\mu_i dn_i \qquad (11\text{-}6)$$

$$dG = -SdT + VdP + Ydy + \Sigma_i\mu_i dn_i \qquad (11\text{-}7)$$

These are the common thermodynamic potential functions. There are other potential functions that can be generated by Legendre transformations: for example, the function $K = K(S, V, Y; n_i)$ and the function $J = J(T, V, Y; n_i)$. Obviously, using Eq. 11-4

$$K = E - y(\partial E/\partial y)_{S,V;\underline{n}_i}$$

$$= TS - PV + \Sigma_i\mu_i n_i \qquad (11\text{-}8)$$

$$J = E - S(\partial E/\partial S)_{V,y;\underline{n}_i} - y(\partial E/\partial y)_{S,V;\underline{n}_i}$$

$$= -PV + \Sigma_i\mu_i n_i \qquad (11\text{-}9)$$

EXERCISES

1. Show that $dK = TdS - PdV - ydY + \Sigma_i \mu_i dn_i$.

2. Show that $dJ = -SdT - PdV - ydY + \Sigma_i \mu_i dn_i$.

Note: If we omit the P-V terms, the K and J functions resemble the enthalpy (H) and the Gibbs free energy (G), except that VdP is replaced by $-ydY$. These functions are sometimes called *elastomer enthalpy* and *free energy* and are even denoted as H and G, respectively, which can be confusing because these symbols are reserved for different Legendre transformations. Such difficulties do not arise with E and A, and we may regard these functions as generalizations of the previously defined potential functions E and A. In the following we will only use E and A potentials.

11.2   P-V, σ𝒜, AND fL WORK

We take n to be fixed. Using $dE = TdS + Xdx$ and $dA = -SdT + Xdx$, we obtain the general relations

$$(\partial E/\partial S)_x = T$$
$$(\partial E/\partial x)_S = X \qquad (11\text{-}10)$$
$$(\partial A/\partial T)_x = -S$$
$$(\partial A/\partial x)_T = X \qquad (11\text{-}11)$$

the Maxwell relation (from dA)

$$-(\partial S/\partial x)_T = (\partial X/\partial T)_x \qquad (11\text{-}12)$$

and the Thermodynamic Equation of State

$$(\partial E/\partial x)_T = T(\partial S/\partial x)_T + X$$
$$= X - T(\partial X/\partial T)_x \qquad (11\text{-}13)$$

We adapt these relations to the various types of works. The results are summarized in Table 11.1.

These relationships are sufficient to enable us to solve simple problems associated with these types of works. However, to apply them, we must know the equations of state of the substances. The simplest forms, the

TABLE 11.1 **Fundamental Thermodynamic Relations in the Treatment of Pressure-Volume (P-V) Work, Surface Enlargement (σA) Work, and Rubber Stretching Work (fL)**

Xdx $\leftrightarrow$ $-$PdV	Xdx $\leftrightarrow$ σdA	Xdx $\leftrightarrow$ f dL
$(\partial E/\partial V)_S = -P$	$(\partial E/\partial A)_S = \sigma$	$(\partial E/\partial L)_S = f$
$(\partial S/\partial V)_T = (\partial P/\partial T)_V$	$(\partial S/\partial A)_T = -(\partial \sigma/\partial T)_A$	$(\partial S/\partial L)_T = -(\partial f/\partial T)_L$
$(\partial E/\partial V)_T = T(\partial P/\partial T)_V - P$	$(\partial E/\partial A)_T = \sigma - T(\partial \sigma/\partial T)_A$	$(\partial E/\partial L)_T = f - T(\partial f/\partial T)_L$

"ideal" equations of state, are summarized below for the three-dimensional (3-D), two-dimensional (2-D), and one-dimensional (1-D) cases.

$$\text{3-D. P-V work}: P\overline{V} = RT \qquad (11\text{-}14)$$

2-D. σ work : $\sigma = \sigma_0(1 - T/T_c)^n$; $\sigma_0 = $ constant and T_c is critical

temperature; $T/T_c < 1$; n is constant $\approx 11/9$ $\qquad (11\text{-}15)$

1-D. fL work : $f = T\phi(L)$; $\phi(L)$is a monotonically increasing

function of L $\qquad (11\text{-}16)$

EXERCISES

1. If a piece of rubber is stretched suddenly (i.e., adiabatically), will it cool, heat up, or stay the same?

SOLUTION

Adiabatically means constant S. Thus,

$$(\partial T/\partial L)_S = -(\partial S/\partial L)_T/(\partial S/\partial T)_L$$
$$= (\partial f/\partial T)_L T/C_L = (T/C_L)\phi(L) > 0 \qquad (11\text{-}17)$$

Thus the rubber heats up.

2. If a liquid surface is increased adiabatically, will it heat up, cool, or stay the same?

SOLUTION

$$(\partial T/\partial \mathcal{A})_S = -(\partial S/\partial \mathcal{A})_T/(\partial S/\partial T)_{\mathcal{A}}$$
$$= (T/C_{\mathcal{A}})(\partial \sigma/\partial T)_{\mathcal{A}}$$
$$= (T/C_{\mathcal{A}})n\sigma_0(1 - T/T_c)^{n-1}(-1/T) < 0 \qquad (11\text{-}18)$$

The surface cools.

CHAPTER 12

PHASE TRANSITIONS AND CRITICAL PHENOMENA

Coexistence of two phases requires that, for "normal" systems, the temperature, pressure, and chemical potential of species i, $\mu_i = (\partial G/\partial n_i)_{T,P;\,n_{j\neq i}}$, have the same values in both phases. No such restrictions are placed on $(\partial G/\partial P)_{T;\,n_i} = V$, $(\partial G/\partial T)_{P;\,n_i} = -S$, or $[\partial G/T)/\partial(1/T)]_{P;\,n_i} = H$.

These first derivatives of the free energy are often discontinuous at the transition of the two phases (i.e., they have different values). Ehrenfest called such transitions *first order*. If these derivatives are continuous, but *their* derivatives are discontinuous (that is, if $(\partial^2 G/\partial P^2)_{T;\,n_i} = \kappa$, $(\partial^2 G/\partial T^2)_{P;\,n_i} = C_P/T$, etc. are discontinuous), Ehrenfest called the transition *second order*. The underlying assumption in the classification of these transitions is that there would be a *jump* in the second-order derivatives between the phases, similar to the jump in the first-order derivatives. But in many substances (perhaps most), the difference between κ and C_P in the two phases is not finite but infinite, resembling a λ-transition. Figure 12.1 show plots of G vs. P and T and the variations of its derivatives V and S in a first-order transition.

Examples of first-order transitions are solid-liquid, liquid-vapor, and allotropic transitions. Examples of second-order transitions are transitions in β-brass (copper-zinc alloy) and NH_4 salts.

Thermodynamics and Introductory Statistical Mechanics, by Bruno Linder
ISBN 0-471-47459-2 © 2004 John Wiley & Sons, Inc.

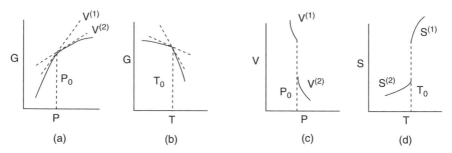

Figure 12.1 Plots of G vs.P, G vs T, V vs. P and S vs. T.

Before we discuss phase transitions, we need to revisit the requirements for stability of the thermodynamic potentials. Focusing on systems in which T, P, and μ_i are uniform throughout "normal" systems, the first derivatives of the thermodynamic potentials with respect to the *extensive* variables are zero and the second derivatives are *positive* except at the critical point, where they are zero. The second derivatives of the potential functions with respect to the *intensive* variables (Section 7.10) are *negative*. The results, as noted before, are often summarized by stating that these derivatives are convex functions of the extensive variables, and concave functions of the intensive variables.

12.1 STABLE, METASTABLE, AND UNSTABLE ISOTHERMS

It is instructive to examine the behavior of the van der Waals equation of state, even though (and perhaps because) the isotherms do not satisfy the stability conditions over the entire range. A typical isotherm of a one-component system is shown in Figure 12.2. We already know that, below the critical point, $(\partial P/\partial V)_T$ has to be negative for stable equilibrium. Thus, the portion BCD is never realized. If the volume starting at G is decreased, the pressure increases to P_F. Further decrease in volume causes the system to split into two phases. The pressure and chemical potentials remain constant until the volume V_E is reached. Thereafter, the system reverts to a one-phase system, and the pressure and chemical potentials increase with further decreases in volume.

The segments EB and DF can be reached on rare occasions. They do not violate the stability rules but represent metastable states (DF—a supersaturated vapor, and EB—an overexpanded liquid). Such states may exist for a short time but if disturbed will quickly convert to the two-state system because there the chemical potential is lower.

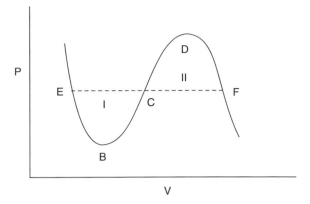

Figure 12.2a   Schematic diagram of a plot of P vs. V for a van der Waals system.

Note that, because the chemical potential of the vapor and liquid must be equal in the coexistence region, i.e., $\mu^G = \mu^L$, the areas I and II in Figure 12.2a, must be equal. To prove this, note that $d\mu = -S\,dT + V\,dP$ and thus, at constant T, the integral from F to E must be zero. Accordingly

$$0 = \int_F^D V dP + \int_D^C V dP + \int_C^B V dP + \int_B^E V dP \qquad (12\text{-}1a)$$

$$= \left(\int_F^D V dP - \int_C^D V dP \right)_I - \left(\int_B^C V dP - \int_B^E V dP \right)_{II} \qquad (12\text{-}1b)$$

The quantities within each set of parentheses in *Eq. 12-1b* represent the values of the areas II and I of Figure 12.2a, respectively, proving that area I = area II.

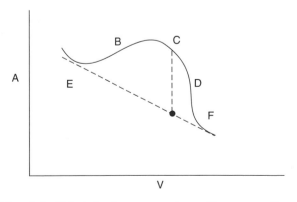

Figure 12.2b   Plot of the Helmholtz free energy, A, vs. V, corresponding to Figure 12.2a.

Figure 12.2b shows the variation of the Helmholtz free energy A with respect to V. corresponding to the van der Waals function of Figure 12.2a. The function A has two minima, one at E and one at F, with common tangents (the same pressure) since $(\partial A/\partial V)_T = -P$. The points E and F represent stable equilibrium. The segments EB and DF represent metastable isotherms, corresponding to supercooled liquid and superheated vapor, respectively. (They are locally but not globally stable.) The points on the dashed line, which correspond to phase separation, have lower A values. The segment BCD is unstable, and a system represented by these points does not exist, at least not for a long time. A system characterized by, say, point C will split into two phases and move to a point on the dashed line, where the free energy is lower.

Similar considerations apply to other thermodynamic potentials. The Gibbs free energy is particularly useful in discussing first-order transitions, since for a one-component system G is proportional to the chemical potential and must have the same value when the two phases coexist.

Although the foregoing discussion was based on the van der Waals system (which is not real), many of its conclusions are applicable to real systems. For example, we know that for any real one-component system, points to the right of the liquid-vapor coexistence curve on a P-T diagram (Figure 12.3) represent pure vapor; to the left of the coexistence line they represent pure liquid. On the coexistence line, a point represents two phases, a liquid phase and a vapor phase, each displaying a minimum on an A vs. V plot (see Fig. 12.4). The dashed curve connecting the two minima represents

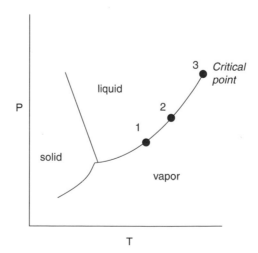

Figure 12.3   First order phase diagram of a one−component system, showing the critical point and two other points on the liquid−vapor coexistence curve.

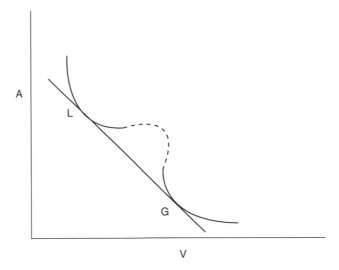

Figure 12.4 Schematic representation of the variation of A vs. V, showing minima corresponding to the points of Figure 12.3.

a continuous, but unstable, fluid. Obviously, such a line must go through a maximum, through a hump.

It is easier to visualize coexisting minima in plots of μ vs. V rather than A vs. V because here the minima are the same level. In a one-component system, μ is just the Gibbs free energy per mole, and the chemical potential at each minimum (having the same pressure and the same temperature) can be obtained from the relation $\mu = \overline{A} - \overline{V}(\partial A/\partial V)_T = A + P\overline{V}$. Obviously, the two coexistence chemical potentials must be equal. At the hump, the chemical potential is larger than the coexistence chemical potentials.

Let X be a point on the liquid-vapor coexistence curve of a P-T phase diagram. The corresponding point on a μ-$\overline{V}$ diagram will show two minima, one at the liquid volume position and on at the gaseous volume position at that temperature and pressure. (*Curve 1* in Figure 12.5 is a schematic representation of μ vs. $\overline{V}$ of point 1 on the P-T diagram in Figure 12.3.) It may be inferred that, when X represents a point to the right of the coexistence curve in Fig. 12.3, there will be only one minimum (the one on the right in Fig. 12.5) and this minimum will be lower than the coexistence minima. When X represents a point to the left of the coexistence curve, the left minimum will be the only minimum and lower than the coexistence minima. Let us imagine that X lies on the left of the coexistence curve and due to fluctuations makes a jump, as a one-phase system, to a higher volume and winds up at the hump. The chemical potential is now higher than at the coexistent minima, and the one-phase system will immediately split into two phases.

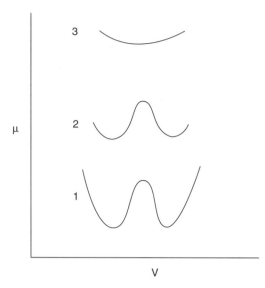

Figure 12.5 Schematic representation of μ vs. V, showing minima corresponding to the points of Figure 12.3.

Note: In a first-order transition, the molar Gibbs free energies of the two phases are equal, but other molar functions, such as S or H, are discontinuous at the transition, as shown in Figure 12.1.

Moving along the vapor-liquid coexistence curve toward the critical point, in Fig. 12.5 the minima get shallower and at the critical point merge into one minimum, which is fairly flat. This has a profound effect on the thermodynamic behavior of the system in the critical region. Beyond the critical point, the minimum becomes steep (very steep) again, and the system behaves normally as a one-phase system.

12.2 THE CRITICAL REGION*

Until about 1960, thermodynamics (i.e., macroscopic thermodynamics) appeared logically complete, except for difficulties associated with critical behavior. Thermodynamics predicted correctly that heat capacities, compressibilities, magnetic susceptibilities, and other second-order derivatives of the free energy diverge; however, it failed to account correctly for the analytic form of the divergence. (Discussions of modern theories of critical

---

*For a comprehensive discussion of phase transitions and critical phenomena, see Stanley, 1971.

behavior are beyond the scope of this book, and we shall only touch on the salient points of the subject.)

It is common practice to describe the behavior of the susceptibilities in the critical region by some "critical exponent," such as α, β, γ, etc. In particular

$$C_V \sim (T - T_c)^{-\alpha} \qquad (T > T_c)$$
$$\sim (T_c - T)^{-\alpha'} \qquad (T < T_c) \qquad (12\text{-}2)$$

$$\kappa \sim (T - T_c)^{-\gamma} \qquad (T > T_c)$$
$$\sim (T_c - T)^{-\gamma'} \qquad (T_c < T) \qquad (12\text{-}3)$$

where T_c is the critical temperature. Along the coexistence curve, the change in volume varies as

$$\Delta V \sim (T_c - T)^{-\beta} \qquad (T < T_c) \qquad (12\text{-}4)$$

According to present thinking, every phase transition is characterized by an "order parameter." For example, in a liquid-vapor phase change, the order parameter is the molar volume; in a Cu-Zn alloy (β-brass), the order parameter is the factional difference of the occupation numbers of the atoms in the so-called super lattices, etc. Early attempts to account for the power law behavior (e.g., of the Landau theory); (Landau and Lifshitz, 1966) are known as *mean field* theories because they use mean values of the order parameters These theories predicted critical exponents, which differed markedly from the experimental values. For example, the critical exponent β, based on the Landau theory, is $\beta = 0.5$, whereas the experimental value is more like $\beta \approx 0.3$–0.4. Similar discrepancies are observed for the other exponents, such as α, etc. In all cases, the predicted values based on classical thermodynamics appear to be too large.

What seems to be the problem in the treatment of critical phenomena? It is known theoretically and experimentally (from statistical mechanics and from scattering techniques, for example) that fluctuations are always present. The validity of thermodynamics is predicated on the assumption that fluctuations are unimportant, which is certainly true when the system is macroscopic. We do not worry whether a particular thermodynamic variable (such as E or H) has a unique value or fluctuates between several values when the system is macroscopic. We assume and treat the thermodynamic variables as being unique.

We have seen that, even in the near vicinity of a coexistence line, where the two minima are very close to each other, the system will reside in the lowest minimum. Although a jump across the hump that separated the two minima is possible and sometimes occurs, resulting in the formation of a metastable state, such occurrences are rare. The situation changes when the critical region is approached. The minima become more and more shallow, and fluctuations become significant. Properties such as the order parameter at certain microscopic regions may vary significantly from the bulk value. The extent of these microscopic regions is referred to as *correlation length*. They become very large the closer the system is to the critical point and play an important role in modern explanations of critical behavior.

What is the upshot of all this? According to Callen 1985, the root of the problem is that, in the critical region, thermodynamics is inapplicable because of the large fluctuations. In macroscopic systems, we associate only one value with each of the thermodynamic variables (such as E or H), although, in reality, there is a distribution of values due to the fluctuations. But the deviations from the average are so small that, for all practical purposes, there is only one value—the average or most probable value. Critical phenomena are exceptions: fluctuations are no longer negligible, although the system as a whole is macroscopic.

The next chapters present an elementary treatment of basic statistical mechanics. Statistical mechanics is not limited to macroscopic systems but may be applied to systems where fluctuations are not negligible.

PART II

INTRODUCTORY STATISTICAL MECHANICS

CHAPTER 13

PRINCIPLES OF STATISTICAL MECHANICS

13.1 INTRODUCTION

Statistical Mechanics (or Statistical Thermodynamics, as it is often called) is concerned with predicting and as far as possible interpreting the macroscopic properties of a system in terms of the properties of its microscopic constituents (molecules, atoms, electrons, etc).

For example, thermodynamics can interrelate all kinds of macroscopic properties, such as energy, entropy, and so forth, and may ultimately express these quantities in terms of the heat capacity of the material. Thermodynamics, however, cannot predict the heat capacities: statistical mechanics can.

There is another difference. Thermodynamics (meaning macroscopic thermodynamics) is not applicable to small systems ($\sim 10^{12}$ molecules or less) or, as noted in Chapter 12, to large systems in the critical region. In both instances, failure is attributed to large fluctuations, which thermodynamics does not take into account, whereas statistical mechanics does.

How are the microscopic and macroscopic properties related? The former are described in terms of position, momentum, pressure, energy levels, wave functions, and other *mechanical* properties. The latter are described in terms of heat capacities, temperature, entropy, and others—that is, in terms of

Thermodynamics and Introductory Statistical Mechanics, by Bruno Linder
ISBN 0-471-47459-2 © 2004 John Wiley & Sons, Inc.

thermodynamic properties. Until about the mid-nineteenth century, the two seemingly different disciplines were considered to be separate sciences, with no apparent connection between them. Mechanics was associated with names like Newton, LaGrange, and Hamilton and more recently with Schrodinger, Heisenberg, and Dirac. Thermodynamics was associated with names like Carnot, Clausius, Helmholtz, Gibbs, and more recently with Carathéodory, Born, and others. *Statistical mechanics is the branch of science that interconnects these two seemingly different subjects.* But statistical mechanics is not a mere extension of mechanics and thermodynamics. Statistical mechanics has its own laws (postulates) and a distinguished slate of scientists, such as Boltzmann, Gibbs, and Einstein, who are credited with founding the subject.

13.2   PRELIMINARY DISCUSSION—SIMPLE PROBLEM

The following simple (silly) problem is introduced to illustrate with a concrete example what statistical mechanics purports to do, how it does it, and the underlying assumptions on which it is based.

Consider a system composed of three particles (1, 2, and 3) having a fixed volume and a fixed energy, E. Each of the particles can be in any of the particle energy levels, ε_i, shown in Figure 13.1. We take the total energy, E, to be equal to 6 units.

Note: Historically, statistical mechanics was founded on classical mechanics. Particle properties were described in terms of momenta, positions, and similar characteristics and, although as a rule classical mechanics is simpler to use than quantum mechanics, in the case of statistical mechanics it is the other way around. It is much easier to picture a distribution of particles among discrete energy levels than to describe them in terms of velocities momenta, etc. Actually, our treatment will not be based on quantum mechanics. We will only use the language of quantum mechanics.

In the example discussed here, we have for simplicity taken the energy levels to be nondegenerate and equally spaced. Figure 13.2 illustrates how

$$\varepsilon_4 = 4$$
$$\varepsilon_3 = 3$$
$$\varepsilon_2 = 2$$
$$\varepsilon_1 = 1$$

Figure 13.1   Representation of a set of equally spaced energy levels.

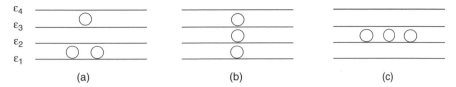

Figure 13.2 Distribution of three particles among the set of energy levels of Figure 13.1, having a total energy of 6 units.

the particles can be distributed among the energy levels under the constraint of total constant energy of 6 units. Although the total energy is the same regardless of how the particles are distributed, it is reasonable to assume that some properties of the system, other than the energy, E, will depend on the arrangement of the particles among the energy states. These arrangements are called *microstates* (or micromolecular states).

Note: It is wrong to picture the energy levels as shelves on which the particles sit. Rather, the particles are continuously colliding, and the microstates continuously change with time.

13.3 TIME AND ENSEMBLE AVERAGES

During the time of measurement on a single system, the system undergoes a large number of changes from one microstate to another. The observed macroscopic properties of the system are *time averages* of the properties of the instantaneous microstates—that is, of the mechanical properties. Time-average calculations are virtually impossible to carry out. A way to get around this difficulty is to replace the time average of a *single* system by an *ensemble* average of a very large collection of systems. That is, instead of looking at one system over a period of time, one looks at a (mental) collection of a large number of systems (all of which are replicas of the system under consideration) at a *given instance of time.* Thus, in an ensemble of systems, all systems have certain properties in common (such as same N, V, E) but differ in their microscopic specifications; that is, they have different microstates. The assumption that the time average may be replaced by an ensemble average is stated as postulate:

- *Postulate I: the observed property of a single system over a period of time is the same as the average over all microstates (taken at an instant of time).*

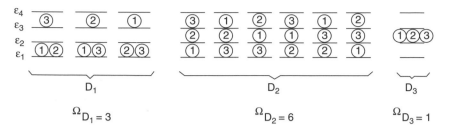

Figure 13.3   Identity of the particles corresponding to the arrangements in Figure 13.2. The symbol Ω_{Di} represents the number of quantum states associated with distribution Di.

13.4   NUMBER OF MICROSTATES, Ω_D, DISTRIBUTIONS D_i

For the system under consideration, we can construct 10 microstates (Figure 13.3). We might characterize these microstates by the symbols Ψ_1, Ψ_2, and so forth. (In quantum mechanics, the Ψ symbols could represent wave functions.) The microstates can be grouped into three different classes, characterized by the particle distributions D_1, D_2, D_3. Let Ω_D, denote the number of microstates belonging to distribution D_1, etc. Thus, $\Omega_{D_1} = 3$, $\Omega_{D_2} = 6$, and $\Omega_{D_3} = 1$.

Each of the systems constituting the ensemble made up of these microstates has the same N, V, and E, as noted before, but other properties may be different, depending on the distribution. Specifically, let χ_1 be a property of the systems when the system is in the distribution D_1, χ_2 when the distribution is D_2, and χ_3 when the distribution is D_3. The ensemble average, which we say is equal to the time average (and thus to the observed property) is

$$\langle\chi\rangle_{ensemble} = \chi_{obs} = (3\chi_1 + 6\chi_2 + \chi_3)/10 \qquad (13\text{-}1)$$

This result is based on a number of assumptions, in addition to the time-average postulate, assumptions that are implied but not stated. In particular

1) *Equation 13-1* assumes that all microstates are equally probable. (Attempts to prove this have been only partially successful.) This assumption is so important that it is adopted as a fundamental postulate of statistical mechanics.
   - *Postulate II: all microstates have equal a priori probability.*
2) Although we refer to the microscopic entities as "particles," we are noncommittal as to the nature of the particles. They can mean elementary particles (electrons, protons, etc.), composites of elementary particles, aggregates of molecules, or even large systems.

3) In this example, the assumption was made that each particle retains its own set of *private* energy level. This is generally not true—interaction between the particles causes changes in the energy levels. Neglecting the interactions holds for ideal gases and ideal solids but not for real systems. In this course, we will only treat ideal systems, and the assumption of private energy levels will be adequate. This assumption is not a necessary requirement of statistical mechanics, and the rigorous treatment of realistic systems is not based on it.

4) In drawing pictures of the 10 microstates, it was assumed that all particles are distinguishable, that is, that they can be labeled. This is true classically, but not quantum mechanically. In quantum mechanics, identical particles (and in our example, the particles are identical) are *indistinguishable*. Thus, instead of there being three different microstates in distribution D_1, there is only one, i.e., $\Omega_{D_1} = 1$. Similarly, $\Omega_{D_2} = 1$ and $\Omega_{D_3} = 1$. Moreover, quantum mechanics may prohibit certain particles (fermions) from occupying the same energy state (think of Pauli's Exclusion Principle), and in such cases distributions D_1 and D_3 are not allowed.

In summary, attention must be paid to the nature of the particles in deciding what statistical count is appropriate.

1) If the particles are classical, i.e., *distinguishable*, we must use a certain type of statistical count, namely the *Maxwell-Boltzmann* statistical count.

2) If the particles are quantal, that is, *indistinguishable* and there are *no restrictions as to the number of particles per energy state*, we have to use the *Bose-Einstein* statistical count.

3) If the particles are quantal, that is, *indistinguishable and restricted to no more than one particle per state*, then we must use the *Fermi-Dirac* statistical count.

4) Although in this book we deal with particles, which for most part are quantal (atoms, molecules, etc), our treatment will not be based on explicit quantum mechanical techniques. Rather, the effects of quantum theory will be taken into account by using the so-called *corrected* classical Maxwell-Boltzmann statistics. This is a simple modification of the Maxwell-Boltzmann statistics but, as will be shown, can be applied to most molecular gases at ordinary temperatures.

5) Although pictures may be drawn to illustrate how the particles are distributed among the energy levels and how the number of microstates can be counted in a given distribution, this can only be

accomplished when the number of particles is small. If the number is large (approaching Avogadro's number), this would be an impossible task. Fortunately, it need not be done. What is important, as will be shown, is knowing the *number of microstates*, Ω_{D^*}, *belonging to the most probable distribution*. There are mathematical techniques for obtaining such information, called *Combinatory Analysis*, to be taken up in Section 13.5.

6) In our illustrative example, the distribution, D_2, is more probable than either D_1 or D_3. Had we used a large number of particles (instead of 3) and a more complex manifold of energy levels, the distribution D_2 would be so much more probable so that, for all practical purposes, the other distributions may be ignored. In terms of the most probable distribution as D*, we can write

$$\langle \chi \rangle = \chi_{obs} = (\Omega_D \chi_1 + \cdots + \Omega_{D^*} \chi^* + \cdots)/\Sigma_D \Omega_{D_i} \approx \chi^* \qquad (13\text{-}2)$$

7) The ensemble constructed in our example—in which all systems have the same N, V, and E—is not unique. It is a particular ensemble, called the *microcanonical* ensemble. There are other ensembles: the *canonical* ensemble, in which all systems have the same N and V but different Es; the *grand canonical ensemble*, in which the systems have the same V but different Es and Ns; and still other kinds of ensembles. Different ensembles allow different kinds of fluctuations. (For example, in the canonical ensemble, there can be no fluctuations in N because N is fixed, but in the grand canonical ensemble, there are fluctuations in N.) Ensemble differences are significant when the systems are small; in large systems, however, the fluctuations become insignificant with the possible exception of the critical region, and all ensembles give essentially the same results. In this course, we use only the microcanonical ensemble.

13.5 MATHEMATICAL INTERLUDE VI: COMBINATORY ANALYSIS

1. In how many ways can N distinguishable objects be placed in N positions? Or in how many ways can N objects be permuted, N at a time?
   - *Result*: the first position can be filled by any of the N objects, the second by $N - 1$, and so forth; thus

$$\text{и} = P_N^N = (N - 1)(N - 2) \ldots 1 = N! \qquad (13\text{-}3)$$

2. In how many ways can m objects be drawn out of N? Or in how many ways can N objects be permuted m at a time?

   - *Result*: the first object can be drawn in N different ways, the second in $N - 1$ ways, and the m*th* in $(N - m + 1)$ ways:

$$И = P_N^m = N(N - 1) \ldots (N - m + 1) \qquad (13\text{-}4a)$$

   - Multiplying numerator and denominator by $(N - m)! = (N - m)(N - m - 1) \ldots 1$ yields

$$И = P_N^m = N!/(N - m)! \qquad (13\text{-}4b)$$

3. In how many ways can m objects be drawn out of N? The identity of the m objects is immaterial. This is the same as asking, In how many ways can N objects, taken m at a time, be *combined?*

   Note: there is a difference between a *permutation* and a *combination*. In a permutation, the identity and order of the objects are important; in a combination, only the identity is important. For example, there are six permutations of the letters A, B, and C but only one combination.

$$И = C_N^m = N!/[(N - m)!m!] \qquad (13\text{-}5)$$

4. In how many ways can N objects be divided into two piles, one containing $N - m$ objects and the other m objects? The order of the objects in each pile is unimportant.

   - *Result*: we need to divide the result given by *Eq. 13-4b* by m! to correct for the ordering of the m objects:

$$И = P_N^m = N!/[(N - m)!m!] \qquad (13\text{-}6)$$

   (This is the same as *Eq. 13-5*.)

5. In how many ways can N (distinguishable) objects be partitions into c classes, such that there be N_1 objects in class 1, N_2 objects in class 2, and so on, with the stipulation that the order within each class is unimportant?

   - *Result*: obviously

$$И = N!\left(\frac{N!}{N_1!N_2!\cdots N_i!}\right) \qquad (13\text{-}7)$$

This expression is the same as the coefficient of multinomial expansion:

$$(f_1 + f_2 + \cdots f_c)^N = \sum \{[N!/(N_1!N_2!\cdots N_C!)]f_1^{N_1}f_2^{N_2}\cdots f_C^{N_i}\} \qquad (13\text{-}8)$$

6. In how many ways can one arrange N distinguishable objects among g boxes. There are no restrictions as to the number of objects per box.
   - *Result*: the first object can go into any of the g boxes, so can the second, and so forth:

$$и = g^N \qquad (13\text{-}9)$$

7. In how many ways can N *distinguishable* objects be distributed into g boxes (g ≥ N) with the stipulation that no box may contain more than one object?
   - *Result*:

$$и = g!/(g - N)! \qquad (13\text{-}10)$$

8. In how many ways can N *indistinguishable* objects be put in g boxes such that there would be no more than one object per box?
   - *Result*:

$$и = g!/[(g - N)!N!] \qquad (13\text{-}11)$$

9. In how many ways can N *indistinguishable* objects be distributed among g boxes? There are no restrictions as to the number of objects per box. Partition the space into g compartments. If there are g compartments, there are g − 1 partitions. To start, treat the objects and partitions on the same footing. In other words, permute N + g − 1 entities. Now correct for the fact that permuting objects among themselves gives nothing new, and permuting partitions among themselves does not give anything different.
   - *Result*:

$$и = (g + N - 1)!/[(g - 1)!N!] \qquad (13\text{-}12)$$

This formula was first derived by Einstein.

13.6 FUNDAMENTAL PROBLEM IN STATISTICAL MECHANICS

We present a set of energy levels $\varepsilon_1, \varepsilon_2, \ldots \varepsilon_i \ldots$; with degeneracies $g_1, g_2, \ldots g_i \ldots$; and occupation numbers $N_1, N_2, \ldots N_i \ldots$. In how many ways can those N particles be distributed among the set of energy levels, with the stipulation that there be N_1 particles in level 1, N_2 particles in level 2, and so forth?

Obviously, the answer will depend on whether the particles are distinguishable or indistinguishable, whether there are restrictions as to how may particles may occupy a given energy state, etc. In this book, we will treat, in some detail, the statistical mechanics of *distinguishable* particles, as noted before, and correct for the indistinguishability by a simple device. The justification for this procedure is given below.

13.7 MAXWELL-BOLTZMANN, FERMI-DIRAC, BOSE-EINSTEIN STATISTICS. "CORRECTED" MAXWELL-BOLTZMANN STATISTICS

13.7.1 Maxwell-Boltzmann Statistics

Particles are distinguishable, and there are no restrictions as to the number of particles in any given state.

Using Combinatory Analysis *Eqs. 13-7, 13-8, 13-9* gives the number of microstates in the distribution, D.

$$\Omega_D^{MB} = [N!/(N_1!N_2!\cdots N_i!\cdots)]g_1^N g_2^N \ldots g_i^N \ldots \qquad (13\text{-}13)$$

13.7.2 Fermi-Dirac Statistics

Particles are indistinguishable and restricted to no more than one particle per state.

Using *Eq. 13.11* of the Combinatory Analysis gives

$$\Omega_D^{FD} = \{g_1!/[(g_1 - N_1)!N_1!]\}\{g_2!/[g_2 - N_2)!N_2!]\} \ldots$$
$$= \Pi_i\{g_i!/(g_i - N_i)!N_i!\} \qquad (13\text{-}14)$$

13.7.3 Bose-Einstein Statistics

Particles are indistinguishable, and there are no restrictions.

Using *Eq. 13-12*, gives

$$\Omega_D^{BE} = [(g_1 + N_1 - 1)!/(g_1 - 1)!N_1!][(g_2 + N_2 - 1)!/(g_2 - 1)!N_2!] \ldots$$
$$= \Pi_i[(g_i + N_i - 1)!/(g_i - 1)!N_i!] \qquad (13\text{-}15)$$

The different statistical counts produce thermodynamic values, which are vastly different. Strictly speaking, all identical quantum-mechanical particles are indistinguishable, and we ought to use only Fermi-Dirac or Bose-Einstein statistics. For electrons, Fermi-Dirac statistics must be used; for liquid Helium II (consisting of He^4) at very low temperature, Bose-Einstein Statistics has to be used. Fortunately, for most molecular systems (except systems at very low temperatures), the number of degeneracies of a quantum state far exceeds the number of particles of that state. For most excited levels $g_i \gg N_i$ and as a result, the Bose-Einstein and Fermi-Dirac Ω values approach a common value, the common value being The Maxwell-Boltzmann Ω_D divided by N!

Proof of the above statement is based on three approximations, all reasonable, when $g_i \gg N_i$. They are

1) Stirling's Approximation

$$\ln N! \approx N\ln N - N \qquad \text{(N large)} \qquad (13\text{-}16)$$

2) Logarithmic expansion, $\ln(1 \pm x) \approx \pm x$ (x small) $(13\text{-}17a)$
3) Neglect of 1 compared with g_i/N_i $(13\text{-}17b)$

EXERCISE

1. Using these approximations show that

$$\ln \Omega_D^{FD} = \ln \Omega_D^{BE} = \Sigma_i N_i[\ln(g_i/N_i) + 1] \qquad (13\text{-}18)$$

2. Also, show that

$$\ln \Omega_D^{MB} = \ln N! + \Sigma_i N_i[1 + \ln(g_i/N_i)] \qquad (13\text{-}19a)$$

$$= N\ln N + \Sigma_i N_i \ln(g_i/N_i) \qquad (13\text{-}19b)$$

which is the same as *Equation 13-18* except for the addition of $\ln N!$.

13.7.4 "Corrected" Maxwell-Boltzmann Statistics

It is seen that Ω_D^{FD} and Ω_D^{BE} reach a common value, namely, $\Omega_D^{MB}/N!$, which will be referred to as *Corrected Maxwell-Boltzmann*. Thus

$$\Omega_D^{CMB} = \Omega_D^{MB}/N! \qquad (13\text{-}20a)$$

or, using Eq. 13-16

$$\ln \Omega_D^{CMB} = \Sigma_i N_i \ln(g_i/N_i) + N \qquad (13\text{-}20b)$$

13.8   SYSTEMS OF DISTINGUISHABLE (LOCALIZED) AND INDISTINGUISHABLE (NONLOCALIZED) PARTICLES

We mentioned in the preceding paragraph that in this course we would be dealing exclusively with molecular systems that are quantum mechanical in nature and therefore will use CMB statistics. Is there ever any justification for using MB statistics? Yes—when dealing with crystalline solids. Although the particles (atoms) in a crystalline sold are strictly indistinguishable, they are in fact localized at lattice points. Thus, by labeling the lattice points, we label the particles, making them effectively *distinguishable*. In summary, both the Maxwell-Boltzmann and the Corrected Maxwell-Boltzmann Statistics will be used in this course, the former in applications to *crystalline solids* and the latter in applications to *gases*.

13.9   MAXIMIZING Ω_D

Let D^* be the distribution for which Ω_{D^*} or rather $\ln \Omega_{D^*}$ is a maximum, characterized by the set of occupation numbers $N_1^*, N_2^*, \ldots N_i^* \ldots$ etc. Although the N_i values are strictly speaking discrete, they are so large that we may treat them as continuous variables and apply ordinary mathematical techniques to obtain their maximum values. Furthermore, because we will be concerned here with the most probable values, we will drop the * designation, from here on, keeping in mind that in the future Ω_D will describe the most probable value. To find the maximum values of N_i, we must have

$$\Sigma_i (\partial \ln \Omega_D / \partial N_i) \delta N_i = 0 \qquad (13\text{-}21)$$

subject to the constraints

$$N \text{ is constant or } \Sigma_i \ \delta N_i = 0 \qquad (13\text{-}22)$$

$$E \text{ is constant or } \Sigma_i \ \varepsilon_i \delta N_i = 0 \qquad (13\text{-}23)$$

If there were no constraints, the solution to this problem would be trivial. With the constraints, not all of the variables are independent. An easy way to get around this difficulty is to use the Method of Lagrangian (or Undetermined) Multipliers. Multiplying *Eq. 13-22* by α and *Equation 13-23* by β and subtracting them from *Equation 13-21* gives

$$\Sigma_i (\partial \ln \Omega_D / \partial N_i - \alpha - \beta \varepsilon_i) \delta N_i = 0 \qquad (13\text{-}24)$$

The Lagrange multipliers make all variables $N_1, N_2, \ldots N_i, \ldots$ effectively independent. To see this, let us regard N_1 and N_2 as the dependent variables and all the other N values as independent variables. Independent means that we can vary them any way we want to or not vary them at all. We choose not to vary $N_4, N_5 \ldots$, etc., that is, we set $\delta N_4, \delta N_5, \ldots$ equal to zero. *Equation 13-24* then becomes,

$$(\partial \ln \Omega_D / \partial N_1 - \alpha - \beta \varepsilon_1) \delta N_1 + (\partial \ln \Omega_D / \partial N_2 - \alpha - \beta \varepsilon_2) \delta N_2$$
$$+ (\partial \ln \Omega_D / \partial N_3) - \alpha - \beta \varepsilon_3) \delta N_3 = 0 \qquad (13\text{-}25)$$

We can choose α and β so as to make two terms zero, then the third term will be zero also. Repeating this process with δN_4, δN_5, etc. shows that for every arbitrary *i* (including subscripts $i = 1$, $i = 2$)

$$\partial \ln \Omega_D / \partial N_i - \alpha - \beta \varepsilon_i = 0 \text{ all } i \qquad (13\text{-}26)$$

13.10 PROBABILITY OF A QUANTUM STATE: THE PARTITION FUNCTION

13.10.1 Maxwell-Boltzmann Statistics

Using *Eq. 13-19b* we first write

$$\ln \Omega_D = (N_1 + N_2 + \cdots N_i \ldots) \ln(N_1 + N_2 + \cdots N_i + \cdots)$$
$$+ (N_1 \ln g_1 + N_2 \ln g_2 + \cdots N_i \ln g_i + \cdots)$$
$$- (N_1 \ln N_1 + N_2 \ln N_2 + \cdots N_i \ln N_i + \cdots) \qquad (13\text{-}27)$$

We differentiate with respect to N_i, which we regard here as particular variable, holding constant all other variables. This gives

$$\partial \ln \Omega_D^{MB} / \partial N_i = \ln N + N/N + \ln g_i - \ln N_i - N_i/N_i$$
$$= \ln(N g_i / N_i) = \alpha + \beta \varepsilon_i \qquad (13\text{-}28)$$

or the probability, $\mathcal{P}_i$, that the particle is in state *i*

$$\mathcal{P}_i = N_i/N = g_i e^{-\alpha} e^{-\beta \varepsilon_i} \qquad (13\text{-}29)$$

It is easy to eliminate $e^{-\alpha}$, since

$$\Sigma_i N_i/N = 1 = e^{-\alpha} \Sigma_i g_i e^{-\beta \varepsilon_i} \qquad (13\text{-}30)$$

or,

$$e^{-\alpha} = 1/(\Sigma_i g_i \varepsilon^{-\beta \varepsilon_i}) \qquad (13\text{-}31)$$

and so,

$$\mathcal{P}_i = N_i/N = g_i e^{-\beta \varepsilon_i}/\Sigma_i g_i e^{-\beta \varepsilon_i} \qquad (13\text{-}32)$$

The quantity in the denominator, denoted as q,

$$q = \Sigma_i g_i e^{-\beta \varepsilon_i} \qquad (13\text{-}33)$$

is called the *partition function*. The partition function plays an important role in statistical mechanics (as we shall see): all thermodynamic properties can be derived from it.

13.10.2 Corrected Maxwell-Boltzmann Statistics

$$\ln \Omega_D^{CMB} = \Sigma_i N_i (\ln g_i - \ln N_i + 1) \qquad (13\text{-}34)$$

$$\partial \ln \Omega_D^{CMB}/\partial N_i = \ln g_i - \ln N_i - N_i/N_i + 1 = \alpha + \beta \varepsilon_i \qquad (13\text{-}35)$$

$$\ln(g_i/N_i) = \alpha + \beta \varepsilon_i \qquad (13\text{-}36)$$

and the probability, $\mathcal{P}_i$, is

$$\mathcal{P}_i = N_i/N = (g_i e^{-\alpha} e^{-\beta \varepsilon_i})/N \qquad (13\text{-}37)$$

Using $\Sigma_i \mathcal{P}_i = 1$, gives

$$e^{-\alpha} = N/(\Sigma_i g_i e^{-\beta \varepsilon_i}) \qquad (13\text{-}38)$$

Finally,

$$\mathcal{P}_i = N_i/N = g_i e^{-\beta \varepsilon_i}/\left(\Sigma_i g_i e^{-\beta \varepsilon_i}\right) \qquad (13\text{-}39)$$

It is curious that the probability of a state, $\mathcal{P}_i$, is the same for the Maxwell-Boltzmann as for the Corrected Maxwell-Boltzmann expression. This is also true for some other properties, such as the energy (as will be shown shortly), but not all properties. The entropies, for example, differ.

The average value of a given property, χ (including the average energy, ε), is for both types of statistics.

$$\langle \chi \rangle = \Sigma_i \chi_i \mathcal{P}_i = \Sigma_i \chi_i g_i e^{-\beta \varepsilon_i} / \Sigma_i g_i e^{-\beta \varepsilon_i} \qquad (13\text{-}40)$$

Also, the ratio of the population in state j to state i, is, regardless of statistics

$$N_j / N_i = (g_j / g_i) e^{-\beta(\varepsilon_1 - \varepsilon_i)} \qquad (13\text{-}41)$$

CHAPTER 14

THERMODYNAMIC CONNECTION

14.1 ENERGY, HEAT, AND WORK

The total energy for either localized and delocalized particles (solids, and gases) is, using *Eq. 13-32* or *Eq. 13-39*,

$$E = \Sigma_i N_i \varepsilon_i = N(\Sigma_i \varepsilon_i \, e^{-\beta \varepsilon_i} / \Sigma_i g_i \, e^{-\beta \varepsilon_i}) \qquad (14\text{-}1)$$

$$= N(\Sigma_i \varepsilon_i g_i e^{-\beta \varepsilon_i} / q) \qquad (14\text{-}2)$$

It follows immediately, that *Eq. 14-2* can be written

$$E = -N(\partial \ln q / \partial \beta)_V \qquad (14\text{-}3)$$

The subscript, V, is introduced because the differentiation of lnq is under conditions of constant ε_i. Constant volume (particle-in-a box!) ensures that the energy levels will remain constant.

Note: The quantity within parentheses in *Eqs. 14-2* and *14-3* represent also the average particle energy, and the equations may also be written as

$$E = N\langle \varepsilon \rangle \qquad (14\text{-}4)$$

Thermodynamics and Introductory Statistical Mechanics, by Bruno Linder
ISBN 0-471-47459-2 © 2004 John Wiley & Sons, Inc.

Let us now consider heat and work. Let us change the system from a state whose energy is E to a neighboring state whose energy is E'. If E and E' differ infinitesimally, we may write for a closed system (N fixed)

$$dE = \Sigma_i \varepsilon_i \, dN_i + \Sigma_i N_i \, d\varepsilon_i \qquad (14\text{-}5)$$

Thus, there are two ways to change the energy: (1) by changing the energy levels and (2) by reshuffling the particles among the energy levels. Changing the energy levels requires changing the volume, and it makes sense to associate this process with *work*. The particle reshuffling term must then be associated with *heat*. In *short*, we define the elements of heat and of work as

$$dq = \Sigma_i \varepsilon_i \, dN_i \qquad (14\text{-}6)$$

$$dw = \Sigma_i N_i \, d\varepsilon_i \qquad (14\text{-}7)$$

14.2 ENTROPY

In our discussion of thermodynamics, we frequently made use of the notion that, if a system is isolated, its entropy is a maximum. An isolated system does not exchange energy or matter with the surroundings; therefore, if a system has constant energy, constant volume, and constant N, it is an isolated system. In statistical mechanics, we noticed that under such constraints the number of microstates tends to a maximum. This strongly suggests that there ought to be a connection between the entropy and the number of microstates, Ω or *thermodynamic probability*, as it is sometimes referred to. But there is a problem! Entropy is additive: the entropy of two systems 1 and 2 is $S = S_1 + S_2$, but the number of microstates of two combined systems is multiplicative, that is, $\Omega = \Omega_1 \times \Omega_2$. On the other hand, the log of $\Omega_1 \times \Omega_2$ is additive. This led Boltzmann to suggest the following (which we will take as a postulate):

- *Postulate III: the entropy of a system is $S = k \ln \Omega$.*

Here, k represents the "Boltzmann constant" (i.e., $k = 1.38066 \times 10^{-23}$ J/K) and Ω refers to the number of microstates, consistent with the macroscopic constraints of constant E, N, and V.

Note: Strictly speaking, the above postulate should include all microstates, that is, $\Sigma_D \Omega_D$, but, as noted before, in the thermodynamic limit, only the most probable distribution will effectively count, and thus we will have the basic definition, $S = k \ln \Omega_{D^*}$.

14.2.1   Entropy of Nonlocalized Systems (Gases)

Using *Eq. 13-34*, we obtain

$$S = k \ln \Omega_D^{CMB} = k \ln \Sigma_i N_i [\ln(g_i/N_i) + 1] \qquad (14\text{-}8)$$

Replacing g_i/N_i by $e^{\beta \varepsilon_i}$ q/N (which follows from *Eq. 13-39*, we get

$$S = k\Sigma_i N_i[\ln(q/N) + \beta\varepsilon_i + 1] \qquad (14\text{-}9)$$

$$= k[N \ln(q/N) + \beta\Sigma_i N_i \varepsilon_i + N] \qquad (14\text{-}10)$$

or

$$S = k(N \ln q + \beta E - N \ln N + N) \qquad (14\text{-}11)$$

14.2.2   Entropy of Localized Systems (Crystalline Solids)

Using *Eq. 13-19b* gives for localized systems

$$S = k \ln \Omega_D^{MB} = k\, N \ln N + k\Sigma_i N_i \ln(g_i/N_i) \qquad (14\text{-}12)$$

Using again *Eq. 13-39* or *Eq. 13-33* to replace g_i/N_i yields

$$S = k[N \ln N + \Sigma_i N_i(\ln(q/N) + \beta\varepsilon)] \qquad (14\text{-}13)$$

$$= k(N \ln N + N \ln q - N \ln N + \beta\Sigma_i N_i \varepsilon_i) \qquad (14\text{-}14)$$

$$= k(N \ln q + \beta E) \qquad (14\text{-}15)$$

14.3   IDENTIFICATION OF β WITH 1/KT

In thermodynamics, heat and entropy are connected by the relation, dS = (1/T) dq_{rev}. We have already identified the statistical-mechanical element of heat, namely, dq = $\Sigma_i \varepsilon_i$ dN_i. Let us now seek to identify dS. Although the entropies for localized and delocalized systems differ, the difference is in N, which for a closed system is constant. Thus, we can treat both entropy forms simultaneously by defining

$$S = k(N \ln q + \beta E + \text{constant})$$

$$= k(N \ln \Sigma_i g_i e^{-\beta\varepsilon_i} + \beta E + \text{constant}) \qquad (14\text{-}16)$$

For fixed N, S is a function of β, V and thus $\varepsilon_1, \varepsilon_2, \ldots, \varepsilon_i, \ldots$. Let us differentiate S with respect to β and ε_i:

$$
\begin{aligned}
dS &= (\partial S/\partial \beta)_{\varepsilon_i}\, d\beta + \Sigma_i (\partial S/\partial \varepsilon_i)_{\beta,\varepsilon,j\neq i}\, d\varepsilon_i \\
&= k[(-N\Sigma_i \varepsilon_i g_i e^{-\beta \varepsilon_i}/\Sigma_i\, g_i e^{-\beta \varepsilon_i})d\beta - N\,\beta(\Sigma_i g_i e^{-\beta \varepsilon_i}/\Sigma_i g_i e^{-\beta \varepsilon_i})d\varepsilon_i \\
&\quad + Ed\beta + \beta dE] \qquad\qquad\qquad\qquad\qquad\qquad\qquad\qquad (14\text{-}17)
\end{aligned}
$$

The first term within brackets of *Eq. 14-17* is $-N\langle\varepsilon\rangle d\beta = -E\,d\beta$ and cancels the third term. The second term of *Eq. 14-17* is (using Eq. 13-39) $-\beta \Sigma_i N_i d\varepsilon_i = -\beta dw$. Therefore,

$$
dS = k\beta(dE - dw) = k\beta dq_{rev} \qquad\qquad (14\text{-}18)
$$

Here dq refers to an element of heat and not to the partition function. The differential dS is an exact differential, since it was obtained by differentiating $S(\beta, \varepsilon_i)$ with respect β and ε_i, and so dq must be reversible, as indicated. Obviously, $k\beta$ must be the inverse temperature, i.e., $k\beta = 1/T$ or

$$
\beta = 1/kT \qquad\qquad (14\text{-}19)
$$

14.4 PRESSURE

From $dw = \Sigma_i N_i\, d\varepsilon_i$, we obtain on replacing N_i (Eq. 13-39)

$$
P = -\partial w/\partial V = -\Sigma_i N_i \partial\varepsilon_i/\partial V \qquad\qquad (14\text{-}20)
$$

$$
= -N\Sigma_i(\partial\varepsilon_i/\partial V)\, g_i\, e^{-\beta \varepsilon_i}/\Sigma_i g_i e^{-\beta \varepsilon_i} \qquad\qquad (14\text{-}21)
$$

Note that the derivative of the logarithm of the partition function, q, is

$$
\partial \ln q/\partial V = \Sigma_i[-\beta(\partial\varepsilon_i/\partial V)g_i(e^{-\beta \varepsilon_i}/\Sigma_i g_i e^{-\beta \varepsilon_i})] \qquad\qquad (14\text{-}22)
$$

Consequently,

$$
P = (N/\beta)(\partial \ln q/\partial V) = NkT(\partial \ln q/\partial V)_T \qquad\qquad (14\text{-}23)
$$

APPLICATION

It will be shown later that the translational partition function of system of independent particles (ideal gases), is

$$q_{tr} = (2\pi mkT/h^2)^{3/2}V \qquad (14\text{-}24)$$

Applying *Eq. 14-23* shows that

$$
\begin{aligned}
P &= NkT\,\partial/\partial V\,[\ln(2\pi mkT/h^2)^{3/2} + \ln V]\\
&= NkT/V
\end{aligned}
\qquad (14\text{-}25)
$$

14.5   THE FUNCTIONS E, H, S, A, G, AND μ

From the expressions of E and S in terms of the partition functions and the standard thermodynamic relations, we can construct all thermodynamic potentials.

1. Energy

$$E = kNT^2(\partial \ln q/\partial T)_V \qquad (14\text{-}26)$$

   This expression is valid for both the localized and delocalized systems.
2. Enthalpy

$$
\begin{aligned}
H &= E + PV\\
&= kNT^2(\partial \ln q/\partial T)_V + kNT(\partial \ln q/\partial V)_T V
\end{aligned}
\qquad (14\text{-}27)
$$

   For an ideal gas, the second term is kNT. For an ideal solid (a solid composed of localized but non-interacting particles), the partition function is independent of volume, and the second term is zero.
3. Entropy
   — *for nonlocalized systems*

$$S = kN[\ln(q/N) + 1] + kNT(\partial \ln q/\partial T)_V \qquad (14\text{-}28)$$

   — *for localized systems*

$$S = kN \ln q + kNT(\partial q/\partial T)_V \qquad (14\text{-}29)$$

4. Helmholtz Free Energy, $A = E - TS$

   — *for nonlocalized systems*

$$A = kNT^2(\partial \ln q/\partial T)_V - kNT^2(\partial \ln q/\partial T)_V$$
$$- kNT[\ln(q/N) + 1] \qquad (14\text{-}30)$$
$$= -kNT\ln(q/N) - kNT \text{ (for ideal gas)} \qquad (14\text{-}31)$$

   — *for localized sytems*

$$A = kNT^2(\partial \ln q/\partial T)_V - kNT\ln q - kNT^2(\partial \ln q/\partial T)_V \qquad (14\text{-}32)$$
$$= -kNT\ln q \text{ (for ideal solid)} \qquad (14\text{-}33)$$

5. Gibbs Free Energy, $G = A + PV$

   — *for nonlocalized systems*

$$G = -kNT\ln(q/N) - kNT + kNT(\partial \ln q/\partial V)_T V \qquad (14\text{-}34)$$
$$= -kNT\ln(q/N) \text{ (for an ideal gas)} \qquad (14\text{-}35)$$

   — *for localized systems*

$$G = -kTN\ln q + kNT(\partial \ln q/\partial V)_T V \qquad (14\text{-}36)$$
$$= -kTN\ln q \text{ (for ideal solid)} \qquad (14\text{-}37)$$

6. Chemical Potential, $\mu = G/N$

In statistical mechanics, unlike thermodynamics, it is customary to define the chemical potential as the free energy per molecule, not per mole. Thus, the symbol μ, used in this part of the course outlined in this book, represent the free energy per molecule.

   — *for nonlocalized systems,*

$$\mu = -kT\ln(q/N) - kT + (kT(\partial \ln q/\partial V)_T)V \qquad (14\text{-}38)$$
$$= -kT\ln(q/N) \text{ (for ideal gas)} \qquad (14\text{-}39)$$

   — *for localized systems*

$$\mu = -kT\ln q + [kT(\partial \ln q/\partial V)_T]V \qquad (14\text{-}40)$$
$$= -kT\ln q \text{ (for an ideal solid)} \qquad (14\text{-}41)$$

Note: Solids, and not only ideal solids, are by and large incompressible. The variation of ln q with V can be expected to be very small (i.e., PV is very small), and no significant errors are made when terms in $(\partial \ln q/\partial V)_T$ are ignored. Accordingly, there is then no essential difference between E and H and between A and G in solids.

We now have formal expressions for determining all the thermodynamic functions of gases and solids. What needs to be done next is to derive expressions for the various kinds of partition functions that are likely to be needed.

CHAPTER 15

MOLECULAR PARTITION FUNCTION

To a very good approximation, the molecular energy may be separated into translational(tr), vibrational(vib), rotational(rot), electronic(el), and even nuclear(nuc) components.

$$\varepsilon \approx \varepsilon_{tr} + \varepsilon_{vib} + \varepsilon_{rot} + \varepsilon_{el} + (\varepsilon_{nuc}) \qquad (15\text{-}1)$$

Because $q = \Sigma_i g_i\, e^{-\beta\varepsilon_i}$, we can write

$$q \approx \Sigma_{tr}\, g_{tr}\, e^{-\beta\varepsilon}\, \Sigma_{vib}\, g_{vib}\, e^{-\beta\varepsilon_i}\, \Sigma_{rot}\, g_{rot}\, e^{-\beta\varepsilon}\, \Sigma_{el}\, g_{el}\, e^{-\beta\varepsilon_i}$$

$$\times \left(\Sigma_{nuc}\, g_{nuc}\, e^{-\beta\varepsilon_i}\right) \qquad (15\text{-}2)$$

$$\approx q_{tr}\, q_{vib}\, q_{rot}\, q_{el} (q_{nuc}) \qquad (15\text{-}3)$$

Note: Although nuclear energy levels are never excited under normal terrestial conditions (we are not talking about accelerators, cyclotrons, etc.), nuclear energy levels can have no direct effect on the thermodynamic properties. They are included here because, as we shall see, nuclear spins have a "veto power" on what rotational energy levels are allowed and thus have a profound effect the rotational partition function.

Thermodynamics and Introductory Statistical Mechanics, by Bruno Linder
ISBN 0-471-47459-2 © 2004 John Wiley & Sons, Inc.

15.1 TRANSLATIONAL PARTITION FUNCTION

We use the particle-in-a-box energy levels to construct the translational partition function. The energy level of a particle in a 3-D box with sides a, b and c is

$$\varepsilon_{n_x,n_y,n_z} = h^2/8m(n_x^2/a^2 + n_y^2/b^2 + n_z^2/c^2) \qquad (15\text{-}4)$$

where n_x, n_y, and n_z are quantum numbers that run from 1 to ∞, m is the mass, and h Planck's constant. Accordingly

$$q_{tr} = \Sigma_{n_x} \exp(-n_x^2 h^2/8ma^2 kT)\Sigma_{n_y} \exp(-n_y^2 h^2/8mb^2 kT)\Sigma_{n_z}$$
$$\times \exp(-n_z^2 h^2/8mc^2/kT) \qquad (15\text{-}5a)$$

Because the spacing between the energy levels is exceedingly small for macroscopic values of a, b, and c we may replace the summations by integrations, i.e.

$$q_{tr} = \int_0^\infty dn_x \exp(-n_x^2 h^2/8makT) \int_0^\infty dn_y \ldots, \int_0^\infty dn_z \ldots \qquad (15\text{-}5b)$$

The integrals are well known. They are of the form $\int_0^\infty dx \exp(-\alpha x^2) = 1/2\sqrt{(\pi/\alpha)}$ and so the integration in *Equation 15-5b* reduces to

$$q_{tr} = (2\pi mkT/h^2)^{3/2}abc = (2\pi mkT/h^2)^{3/2}V \qquad (15\text{-}6)$$

Note: The translational partition function can be used only for gases. In crystalline solids, the particles are located at lattice points and there is no translational energy.

From the translational partition function, we can immediately obtain part of the energy associated with translational motion. For a monatomic gas, the translational energy is the only prevailing energy, except for an occasional contribution from the electronic partition function. Applying *Equation 14-26* gives

$$E_{tr} = kNT^2(\partial \ln q_{tr}/\partial T)_V$$
$$= kNT^2 \partial/\partial T[\ln T^{3/2} + \ln(2\pi mk/h^2)^{3/2}V] \qquad (15\text{-}7)$$
$$= (3/2)\, kNT \qquad (15\text{-}8)$$

The heat capacity is obviously,

$$C_{V,tr} = (\partial E/\partial T)_V \qquad (15\text{-}9)$$

15.2 VIBRATIONAL PARTITION FUNCTION: DIATOMICS

We use the harmonic oscillator as a model for vibrational motion. The energy levels are

$$\varepsilon_v^{vib} = (v + 1/2)h\nu_0 \qquad v = 0, 1, 2, \ldots \qquad (15\text{-}10)$$

where ν_0 is the vibrational frequency of the molecule and $(1/2)h\nu_0$ is the zero-point energy.

$$q_{vib} = \sum_{v=0}^{\infty} e^{-(v+1/2)h\nu_0/kT} = e^{-(1/2)h\nu_0/kT} \sum_{0}^{\infty} e^{-vh\nu_0/kT} \qquad (15\text{-}11a)$$

$$= e^{-(1/2)h\nu_0/kT}(1 + e^{-h\nu_0/kT} + e^{-2h\nu_0/kT} + \cdots) \qquad (15\text{-}11b)$$

$$= e^{-(1/2)h\nu_0/kT}/(1 - e^{-h\nu_0/kT}) \qquad (15\text{-}11c)$$

It is customary to define a "characteristic vibrational temperature," $\theta_v h\nu_0/k$, and thus write

$$q_{vib} = e^{-(1/2)\theta_v/T}/(1 - e^{-\theta_v/T}) \qquad (15\text{-}12)$$

For most diatomic molecules $\theta_v > 300$ K (I_2 and to lesser extent Br_2 are exceptions), and so $q_{vib} \approx 1$ and thus makes no contribution to the thermodynamic properties at ordinary temperatures.

Note: The spacing between vibrational energy levels in diatomic molecules is generally large and the partition function cannot be approximated by the classical limit, which means replacing the summation by integration. If this were permitted the result would be

$$q_{vib,class} = \int_0^{\infty} dv e^{-vh\nu_0/kT} = kT/h\nu_0 = T/\theta_v. \qquad (15\text{-}13)$$

Classically, there is no zero-point energy.

15.3 ROTATIONAL PARTITION FUNCTION: DIATOMICS

The rotational energy of a diatomic molecule is

$$\varepsilon_J = J(J + 1)(h^2/8\pi^2 I) \qquad J = 0, 1, \ldots, \infty \qquad (15\text{-}14)$$

where J is the rotational quantum number; I is the moment of inertia, i.e., $I = \mu d^2$, in which μ is the reduced mass $(1/\mu = 1/m_1 + 1/m_2)$; and d is the distance between the centers of the atoms in the molecule. (The moment of inertia can be obtained spectroscopically from the rotational constant $B = h^2/8\pi^2 Ic$, where c is the speed of light.)

The rotational energy levels are degenerate, $g_J = 2J + 1$. Accordingly

$$q_{rot} = \Sigma_J(2J + 1)\exp[-J(J + 1)(h^2/8\pi^2 IkT)] \qquad (15\text{-}15)$$

In analogy to the vibrational temperature, it is convenient to define a "characteristic rotational temperature," $\theta_r = h^2/8\pi^2 Ik$, and so

$$q_{rot} = \Sigma_J(2J + 1)\exp[-(J^2 + J)\theta_r/T] \qquad (15\text{-}16a)$$

In general, the quantity θ_r/T is small. (For example, for H_2 $\theta_r = 85.4$ K; for O_2 it is 2.07 K.) This allows the summation to be replaced by an integration

$$q_{rot} = \int_0^\infty dJ(2J + 1)\exp[-(J^2 + J)\theta_r/T] \qquad (15\text{-}16b)$$

If we let $y = J^2 + J$, then $dy = (2J + 1)dJ$ and so

$$q_{rot} = \int_0^\infty dy\, e^{-y\theta_r/T} = T/\theta_r \qquad (15\text{-}17)$$

This expression is valid for heteronuclear diatomic molecules. For homonuclear diatomic molecules, the expression has to be divided by 2. (The reason for this has to do with nuclear spin degeneracy, to be discussed later.) By defining a "symmetry" number σ, which has the significance that $\sigma = 1$ when the molecule is heteronuclear and $\sigma = 2$ when the molecule is homonuclear, we can write a general formula

$$q_{rot} = T/\sigma\theta_r \qquad (15\text{-}18)$$

The rotational energy is

$$E = kNT^2\partial/\partial T[\ln T - \ln(\sigma\theta_r)] = kNT \qquad (15\text{-}19)$$

or $E = RT$ per mole, which is the same for nonlocalized and localized systems. The entropies are different because of the presence of σ.

15.4 ELECTRONIC PARTITION FUNCTION

As a rule, only the ground electronic state needs to be considered, since the excited states are so much higher that, with few exceptions, the probability that they will be populated at normal temperatures is essentially nil. If we take the ground state energy to be zero, i.e., $\varepsilon_{el,0} = 0$

$$q_{el} = g_{el,0} + g_{el,1}e^{-\beta\varepsilon_1} + \cdots \cong g_{el,0} \qquad (15\text{-}20)$$

The ground state degeneracies can be determined from the term symbols (e.g., $^1S_0$, $^3P_2$, $^1\Sigma_g$, $^2\Pi_{3/2}$, etc.) We will give the degeneracies for purposes of this book. It is noteworthy that for most molecules $g_{el,0} = 1$ (O_2 is a notable exception; the term symbol is $^3\Sigma_g$ and the degeneracy is $g_{el,0} = 3$).

Obviously, electronic degeneracies have no effect on the energy, but they may affect the entropy because they are not always unity.

15.5 NUCLEAR SPIN STATES

Nuclear energy levels are million electron volts (MEV) apart and, as noted before, never need be considered in statistical problems at terrestrial temperatures. Nuclei have nuclear degeneracies; however, in chemical transformations, the nuclei do not change, and nuclear degeneracies can be ignored (in contrast to electronic degeneracies). Nonetheless, nuclear spin states play a decisive role in the sense that they exert a "veto power" over the kind of rotational energy levels that are permitted in homonuclear molecules. Specifically, nuclear spins (denoted as I) can combine to give rise to symmetric spin states (called *ortho*) and to antisymmetric spin states (called *para*). (If the nuclear spin is I, there are $(I + 1)(2I + 1)$ *ortho* states and $I(2I + 1)$ *para* states.)

In homonuclear diatomic molecules, the following rules apply:

1) If the nuclear mass number (number of neutrons and protons) of each nucleus is *even*, then odd rotational states ($J = 1, 3$, etc.) must combine with *para* spin states and even rotational states ($J = 0, 2$, etc.) must combine with *ortho* spin states.

2) If the nuclear mass number (number of neutrons and protons) of each nucleus is *odd*, then even rotational states ($J = 0, 2$, etc.) must combine with *para* spin states and odd rotational states ($J = 1, 3$, etc.) must combine with *ortho* spin states.

The net result is that, in homonuclear diatomic molecules (except for H_2 and D_2 at low temperatures), roughly half of the rotational energy levels is eliminated from the summation in the nuclear partition function, yielding $q_{rot} = T/2\theta_{rot}$). Nuclear spin states have no effect on the rotational states of heteronuclear diatomics.

15.6   THE "ZERO" OF ENERGY

From the standpoint of formal theory, it is unimportant to know what the precise values of the energy levels are. In many instances, one takes the lowest energy level to be zero. But there cases (for example, chemical reactions) in which it is imperative that the states of all molecular species be measured from a common reference state, which is taken to be the "zero of energy."

Let the set $(0, \varepsilon_1, \varepsilon_2, \ldots)$ represent energy levels in which the lowest level is taken to zero and the set $(0, \varepsilon'_1, \varepsilon'_2, \ldots)$ represents energy levels that are measured from an arbitrary reference energy (not necessarily the lowest energy level) of the molecule (see Figure 15.1). It should be noted that the reference state (zero of energy) does not have to lie below the nonprimed set but can be anywhere. Defining the partition functions based on the primed set energy levels as q' and the partition function based on the non-primed set as q, we get

$$q' = \Sigma_i g_i e^{-\beta \varepsilon'} \qquad (15\text{-}21)$$

$$q = \Sigma_i g_i e^{-\beta \varepsilon} \qquad (15\text{-}22)$$

primed and non-primed sets as

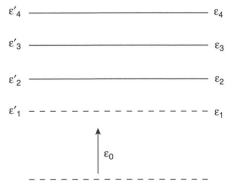

Figure 15.1 Schematic representation of the zero of energy. The primed states are measured from an arbitrary reference state (the zero of energy is indicated by the dashed line). The unprimed states are measured from the lowest molecular state, which lies $\varepsilon 00$ above the zero of energy.

and noticing that

$$\varepsilon_i' = \varepsilon_i + \varepsilon_0 \tag{15-23}$$

gives

$$q' = e^{-\beta\varepsilon_0}\Sigma_i g_i \exp(-\beta\varepsilon_i) \tag{15-24}$$

$$q' = qe^{-\beta\varepsilon_0} \tag{15-25}$$

Accordingly,

$$E' = -N\,\partial/\partial\beta\ln q' = -N\partial\ln q/\partial\beta + N\varepsilon_0$$
$$= kNT^2(\partial\ln q/\partial T)_V + E_0 \tag{15-26}$$

where $E_0 = N\varepsilon_0$ and is constant. Thus, the energy can still be expressed in terms of the ordinary q provided it is denoted by $E' - E_0$. Rather than using E' to represent the energy, it is customary to use the standard notation E but replace it by $E - E_0$. Thus,

$$E - E_0 = kNT^2(\partial\ln q/\partial T)_V \tag{15-27}$$

What effect, if any, does zero of energy have on the entropy? Using again *Eqs. 14-10* and *14-13* we obtain for

1) *Nonlocalized systems:*

$$S = k\Sigma_i N_i[\ln(q'/N) + 1 + \beta\varepsilon_i'] \tag{15-28}$$

$$= k\Sigma_i N_i\{\ln[(q/N)e^{-\beta\varepsilon_i^0}] + 1 + \beta(\varepsilon_i + \varepsilon_0)\} \tag{15-29}$$

$$= k\Sigma_i N_i[\ln(q/N) + 1 + \beta\varepsilon_i] \tag{15-30}$$

$$= kN[\ln(q/N) + 1 + T(\partial\ln q/\partial T)_V] \tag{15-31}$$

2) *Localized systems*

$$S = kN\ln N + k\Sigma_i N_i[\ln(q/N)e^{-\beta\varepsilon_i^0}] + \beta(\varepsilon_i + \varepsilon_0) \tag{15-32}$$

$$= kN\ln N + kN\ln q/N + \Sigma_i N_i\varepsilon_i \tag{15-33}$$

$$= kN[\ln q + T(\partial\ln q/\partial T)_V] \tag{15-34}$$

Comparing *Equations 15-31* with 14-28 and *Equations* 15-34 wtih 14-29 shows that they are identical. Thus, the zero of energy has no effect on the entropy of either localized or delocalized system.

In summary, because the thermodynamic potentials are combinations of S and/or PV, the thermodynamic functions labeled E, H, A, and G in Section 14-5 ought to be replaced by $E - E_0$, $H - E_0$, $A - E_0$, $G - E_0$. The chemical potential should be written (because it represents the free energy per molecule) as $\mu - \varepsilon_0$. The expressions for the entropy remain unchanged.

CHAPTER 16

STATISTICAL MECHANICAL APPLICATIONS

As an illustration of the use of statistical mechanics, four types of applications will be considered here: three having to do with delocalized systems (gases) and one with a localized system (crystalline solids). The simplest applications are calculations of population ratios, thermodynamic functions, equilibrium constants, and heat capacity of solids.

16.1 POPULATION RATIOS

EXERCISE

The vibrational and rotational characteristic temperatures of O_2 are, respectively, $\theta_v = 2228$ K and $\theta_r = 2.050$ K. The ground state electronic degeneracy is $g_{el} = 3$. Calculate the rotational population ratio of O_2 in the state $J = 5$ to $J = 2$ at $T = 300$ K.

Thermodynamics and Introductory Statistical Mechanics, by Bruno Linder
ISBN 0-471-47459-2 © 2004 John Wiley & Sons, Inc.

SOLUTION

$$N_5/N_2 = (2 \times 5 + 1)/(2 \times 2 + 1)\exp[(-5 \times 6 + 2 \times 3)$$

$$\times 2.050/300] \qquad (16\text{-}1)$$

$$= 2.2 \times 0.8487 = 1.86714 \qquad (16\text{-}2)$$

Incidentally, the rotational probability distribution, $\mathcal{P}_J$, goes through a maximum. It is easy to determine the maximum J vale. In general, we require that

$$d\mathcal{P}_J/dJ = 0 \qquad (16\text{-}3)$$

which gives

$$(2/q_{rot})\exp -[(J^2 + J)\theta_r/T] - (1/q_{rot})(2J + 1)^2(\theta_r/T)\exp -[(J^2 + J)\theta_r/T] = 0 \qquad (16\text{-}4)$$

or

$$2 - (2J + 1)^2\theta_r/T = 0 \qquad (16\text{-}5)$$

Thus,

$$J_{max} = (1/2)\left[\sqrt{(2T/\theta_r)} - 1\right] \qquad (16\text{-}6)$$

For O_2, $J_{max} = (1/2)[(2 \times 300/2.050)^{1/2} - 1] = 8.05$ at 300 K and thus the maximum J is around 8.

16.2 THERMODYNAMIC FUNCTIONS OF GASES

EXERCISE

Calculate S and E of 1 mol of O_2 at T = 298.15 K and 1 atm using the data of the example given above.

SOLUTION

The q values are dimensionless. In cgs units $(1 \text{ atm} = 10^6 \text{ dyn/cm}^2)$

$$q_{tr} = (2\pi mkT/h^2)^{3/2}V = (2\pi mkT/h^2)^{3/2}NkT/P \qquad (16\text{-}7)$$
$$= \{[2 \times \pi \times (32.0/6.02 \times 10^{23}) \times 1.38 \times 10^{-16}$$
$$\times 298.15]/(6.62 \times 10^{-27})^2\}^{3/2}$$
$$\times 6.02 \times 10^{23} \times 1.38 \times 10^{-16} \times 298.15/10^6 \qquad (16\text{-}8)$$
$$= 4.35 \times 10^{30} \qquad (16\text{-}9)$$

$$E_{tr} = kNT^2 \frac{\partial}{\partial T}(\ln T^{5/2} + \ln \text{constant}) \qquad (16\text{-}10)$$
$$= (5/2)RT \qquad (16\text{-}11)$$

$R = kN = 1.38 \times 10^{-16} \text{ erg/K molecule} \times 6.02 \times 10^{23} \text{ molecules/mol} = 8.31 \times 10^7 \text{ erg/Kmol} = 8.31 \text{ J/Kmol}$. The translational entropy is

$$S_{tr} = R\{[\ln(q_{er}/N) + 1] + 5/2\} \qquad (16\text{-}12)$$
$$= (8.31 \text{ J/Kmol})\{[\ln(4.35 \times 10^{30}/6.02 \times 10^{23}) + 1] + 5/2\} \qquad (16\text{-}13)$$
$$= 160.32 \text{ J/Kmol} \qquad (16\text{-}14)$$
$$q_{vib} = e^{-1/2(hv_0/kT)}/(1 - e^{-hv_0/kT}) = e^{-(1/2)\theta_v/T}/(1 - e^{-\theta_v/T})$$
$$= e^{-3.714}/(1 - e^{-7.43}) \qquad (16\text{-}15a)$$
$$= e^{-1/2 \times 2228/298.15}/(1 - e^{-2228/298.15}) \qquad (16\text{-}15b)$$
$$= 2.384 \times 10^{-2}/(1 - 5.95 \times 10^{-4}) = 2.384 \times 10^{-2} \qquad (16\text{-}15c)$$
$$E_{vib} = kNT^2(\partial \ln q_{vib}/\partial T) = R[\theta_v/2 + \theta_v/(e^{\theta/T} - 1)] \qquad (16\text{-}16)$$
$$= R[(2228/2) + 2228/(e^{7.43} - 1)] = R(1113.1 + 1.326) \qquad (16\text{-}17)$$
$$= R(1114.4) \qquad (16\text{-}18)$$

This energy is essentially the zero-point energy, confirming that higher vibrational states are not activated. Such energies are often not included in E_{vib} but lumped together with the zero of energy E_0. The vibrational entropy is

$$S_{vib} = R \ln q_{vib} + E_{vib}/T$$
$$= 8.31 \text{ J/Kmol}(\ln 2.384 \times 10^{-2} + 1114.4/298.15) \qquad (16\text{-}19)$$
$$= 8.31 \text{ J/Kmol}(-3.73 + 3.74) = 0 \qquad (16\text{-}20)$$

Note 1: The vibrational entropy is basically zero, which is not surprising because the vibrational energy is essentially the zero-point energy whose contribution is cancelled by the contribution from q_{vib}.

Note 2: Division of the logarithmic expression of q by N came about as a result of making the distinguishable Maxwell-Boltzmann system applicable to the indindistinguishable Corrected Maxwell-Boltzmann system, applicable to delocalized systems. Division by N should be applied only once. It is natural to apply it to the translational entropy as in Eq. 16-12

The rotational contribution is essentially classical

$$q_{rot} = T/\sigma\theta_{rot} = 298.15/2 \times 2.05 = 72.71 \qquad (16\text{-}21)$$
$$E_{rot} = RT \qquad (16\text{-}22)$$
$$S_{rot} = R \ln q_{rot} + E_{rot}/T = 8.31 \text{ J/Kmol}(\ln 72.71 + 1) \qquad (16\text{-}23)$$
$$= 43.96 \text{ J/Kmol} \qquad (16\text{-}24)$$

It should be noted that the rotation entropy takes account of the symmetry number, a reflection of the nuclear spin states.

The electronic contribution is

$$q_{el} = 3 \qquad (16\text{-}25)$$
$$E_{el} = 0 \qquad (16\text{-}26)$$
$$S_{el} = R \ln q_{el} = 8.31 \text{ J/Kmol} \ln 3 = 9.13 \text{ J/Kmol} \qquad (16\text{-}27)$$

Note: O_2 is unusual in the sense that electronic ground state is degenerate. Virtually all homonuclear diatomics are non-degenerate, and thus there are no contributions made to the entropy from electronic states. The total entropy of oxygen at 300K and 1 atm. is the sum of the above entropies.

$$S = S_{tr} + S_{vib} + S_{rot} + S_{el} \qquad (16\text{-}28)$$
$$= 213.41 \text{ J/Kmol} \qquad (16\text{-}29)$$

16.3 EQUILIBRIUM CONSTANTS

Three types of equilibrium constants will be considered:

1) K_N, which expresses the equilibrium constant in terms of the ratio of the number of molecules N_i;
2) K_c, which expresses the equilibrium constant in terms of molecular concentrations N_i/V; and
3) K_P, which expresses the equilibrium constant in terms of partial pressures $P_i = kTN_i/V = kTc_i$.

In statistical mechanics, concentrations are most often expressed as number of molecules per unit volume (rather than moles per unit volume). Recall from thermodynamics that a system is in equilibrium when $\Sigma_i v_i \mu_i = 0$. Here the v_i are the stoichiometric coefficients of the products (assigned positive values) and of the reactants (assigned negative values). Since for gases (see *Eqs. 15-25* and *14-39*)

$$\mu_i = -kT \ln q_i/N_i + \varepsilon_{i,0} \qquad (16\text{-}30)$$

$$\Sigma_i v_i \mu_i = -kT\Sigma_i v_i \ln q_i/N_i + \Sigma_i v_i \varepsilon_{i,0} = 0 \qquad (16\text{-}31)$$

Writing $\Sigma_i v_i \varepsilon_{i,0} = \Delta\varepsilon_0$ gives

$$\ln[\Pi_i(q_i/N_i)^{v_i} e^{-\Delta\varepsilon_0/kT}] = 0 \qquad (16\text{-}32)$$

$$\Pi_i(q_i/N_i)^{v_i} e^{-\Delta\varepsilon_0/kT} = 1 \qquad (16\text{-}33)$$

$$\Pi_i q_i^{v_i} e^{-\Delta\varepsilon_0/kT} = \Pi_i N_i^{v_i} = K_N \qquad (16\text{-}34)$$

$$K_c = \Pi_i(N_i/V)^{v_i} = \Pi_i(q_i/V)^{v_i} e^{-\Delta\varepsilon_0/kT} \qquad (16\text{-}35)$$

K_p is defined in terms of the partial pressure P_i, namely,

$$K_p = \Pi_i P_i^{v_i}$$

Replacing P_i by $(N_i/V)kT$ yields

$$K_P = (kT)^{\Delta v}\Pi_i(q_i/V)^{v_i} e^{-\Delta\varepsilon_0/kT} \qquad (16\text{-}36)$$

In equilibrium problems involving dissociation of molecules, it is customary to take the reference state as the state of the dissociated molecule. Figure 16.1 shows a typical potential energy diagram of a diatomic molecule. Depicted in the diagram are the dissociation energy D_0, and the potential well energy D_c. The symbol ε_0 stands for the energy difference

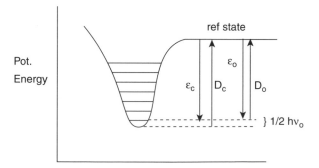

Figure 16.1   Potential energy plot of a diatomic molecule, showing relations between the ε_0 and ε_c. D_C is the energy difference between lowest molecular state and the reference state and D_0 is the dissociation energy.

between the state $v = 0$, and the reference state; ε_c is the energy difference between the potential well and the reference state. Obviously, $D_0 = -\varepsilon_0$, $D_c = -\varepsilon_c$. Moreover, $D_c = D_0 + (1/2)hv_0$. The difference between D_c and D_0 is the zero-point energy.

Note: There are two ways to solve the equilibrium problem, by using D_e and the vibrational partition function q_{vib}, which includes contributions from the zero-point energy (as we have done) or by using D_0 and a vibrational partition function $q_{vib,0}$, which contains no contributions from the zero-point energy. The results are identical. Because the dissociation energies are measured directly, it is best to use D_0 and q_{vib} without the zero-point energy.

EXERCISE

Calculate the equilibrium constant K_P at $T = 1,000$ K for the reaction

$$2Na \rightleftharpoons Na_2 \qquad (16\text{-}37)$$

The dissociation energy for Na_2 is $D_0 = 1.20 \times 10^{-12}$ erg/molecule (i.e., 17.3 kcal/mol). For Na_2, $\theta_v = 229$ K, $\theta_r = 0.221$ K, and $g_{el,0} = 1$. For Na, $g_{el,0} = 2$.

SOLUTION

$$K_P = 1/kT(q_{Na_2}/V)/(q_{Na}/V)^2 \qquad (16\text{-}38a)$$

$$= 1/kT\{[(q_{Na_2,tr}/V)q_{rot}q_{vib}q_{Na_2,el}]/[(q_{Na,tr}/V)^2 q_{Na,el^2}]\}$$

$$\times e^{-(\varepsilon_0 - 0 - 0)/kT} \qquad (16\text{-}38b)$$

Replacing ε_0 by $-D_0$ and omitting the zero-point energy in q_{vib}, we obtain (in cgs units)

$$1/kT = 1/(1.38 \times 10^{-16} \times 1,000) \qquad (16\text{-}39)$$

$$q_{Na_2}/V = [2\pi \times (46/6.02 \times 10^{23}) \times 1.38 \times 10^{-16}$$

$$\times 1,000/(6.63 \times 10^{-27})^2]^{3/2} \times 1,000/(2 \times 0.221)$$

$$\times 1/(1 - e^{-229}/1,000) \times \exp[1.20 \times 10^{-12}/(1.38$$

$$\times 10^{-16} \times 1,000)] \qquad (16\text{-}40)$$

$$(q_{Na}/V)^2 = \{[2\pi \times (23/6.02 \times 10^{23}) \times 1.38 \times 10^{-16}$$

$$\times 1,000/(6.63 \times 10^{-27})^2]^{3/2} \times 2\}^2 \qquad (16\text{-}41)$$

Note: The q values are dimensionless (why?), and so q/V has the dimensions of V^{-1} and K_P has the dimensions of V/kT, which is pressure in cgs units $(dyn/cm^2)^{-1}$. The result of this calculation is $K_P = 0.50 \times 10^{-6}$ $(dyn/cm^2)^{-1}$ or 0.50 atm^{-1}. (The experimental value is $K_P = 0.475$ atm^{-1}.)

16.4 SYSTEMS OF LOCALIZED PARTICLES: THE EINSTEIN SOLID

The Einstein model is a crystalline solid in which each atom vibrates in three dimensions independently from one another. Einstein formulated the theory in 1907, using the quantal expression for the vibration energy of an oscillator, namely, $\varepsilon_v = (v + 1/2)h\nu_0$. This produced a heat capacity curve that went to zero as $T \to 0$, in agreement with observations. Classical statistical mechanics predicted a heat capacity that was constant over the entire range of temperatures, including absolute zero. (The heat capacity of solids, the photo-electric effect, and black-body radiation were three experimental observations that defied classical explanation around the turn of the 20th century, leading eventually to the demise of classical mechanics on the molecular level).

Consider a crystalline solid consisting of 3N linear harmonic oscillators. (We could equally well have picked a system composed of N three-dimensional harmonic oscillators.) For a single harmonic oscillator, the partition function is

$$q_{vib}^1 = e^{-(1/2)h\nu_0/kT}/(1 - e^{-h\nu_0/kT}) \qquad (16\text{-}42)$$

For 3N independent harmonic oscillators

$$q_{vib}^{3N} = [e^{-(1/2)Nh\nu_0/kT}/(1 - e^{-h\nu_0/kT})]^{3N} \qquad (16\text{-}43)$$

This result produces energy, heat capacity and entropy of the solid expressions, detailed in the following sections.

16.4.1 Energy

$$E_{vib} - E_0 = kT^2 \frac{\partial}{\partial T} \ln q_{vib}^{3N} \qquad (16\text{-}44a)$$

$$= (3N/2)h\nu_0 + 3NkT^2[(h\nu_0/kT^2)e^{-h\nu_0/kT}/(1 - e^{(-h\nu_0/kT)})] \qquad (16\text{-}44b)$$

$$= (3/2)Nh\nu_0 + 3Nh\nu_0/(e^{h\nu_0/kT} - 1) \qquad (16\text{-}44c)$$

(The constant zero-point energy is frequently lumped together with the zero of energy term, E_0.)

16.4.2 Heat Capacity

$$C_V = dE_{vib}/dT = 3Nhv_0[-e^{hv_0/kT}(-hv_0/kT^2)]/(e^{hv_0/kT} - 1)^2 \qquad (16\text{-}45a)$$

$$= 3Nk(hv_0/kT)^2 e^{hv_0/kT}/(e^{hv_0/kT} - 1)^2 \qquad (16\text{-}45b)$$

16.4.3 Entropy

$$S = 3kN \ln[e^{-(1/2)hv_0/kT}/(1 - e^{-hv_0/kT})] + E_{vib}/T \qquad (16\text{-}46a)$$

$$= 3kN \ln(1 - e^{-hv_0/kT})^{-1} - (3/2)kNhv_0/kT$$

$$+ (3/2)Nhv_0/T + 3Nhv_0/T(e^{-hv_0/kT} - 1) \qquad (16\text{-}46b)$$

$$= -3kN \ln(1 - e^{-hv_0/kT}) + 3Nhv_0/T(e^{hv_0/kT} - 1) \qquad (16\text{-}46c)$$

Note: In the classical limit, $kT \gg hv_0$ and

$$E - E_0 \approx 3Nhv_0/(1 + hv_0/kT + \cdots - 1) \approx 3kNT = 3RT/mol \qquad (16\text{-}47)$$

and the heat capacity is

$$C_V = 3kN = 3R/mol \qquad (16\text{-}48)$$

The latter agrees with the empirical value of Dulong and Petit.

Note: Figure 16.2 gives a schematic diagram of the variation of the molar heat capacity of a crystalline solid with temperature. The Einstein theory predicts correctly the gross features of the heat capacity curve, namely, that it approaches 3R at high temperature and goes to zero at $T = 0$ K. It does not agree quantitatively with the experimental curve at low temperatures, primarily because the model is oversimplified.

Debye in 1912 used a more realistic model, in which the oscillators are not completely independent but coupled and obtained results that are better at low temperatures resulting in the Debye Cube Law, $C_V \approx aT^3$, which we used.

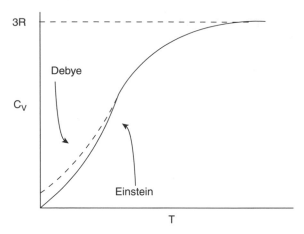

Figure 16.2 Schematic plot of C_V vs. T based on the Einstein Theory. The dotted curve is a correction based on the Debye theory.

16.5 SUMMARY

In summary, the statistical mechanical treatment presented here is based on the notion that each molecule (particle) has its own set of "private" energy levels and that these energy levels are unaffected by the presence of others. This was done on purpose to keep the treatment as simple as possible in an introductory course. As a consequence, the theory was confined to *ideal systems* and only applications to ideal gases and ideal solids were presented to exemplify the use of statistical mechanics. Statistical mechanics is, of course, not limited to such systems and, in fact, has a wide range of applications, including liquids, solids, nonideal gases, electrons, magnetic systems, and a host of other phenomena. However, their treatments requires more sophisticated techniques, techniques that are beyond the scope of this course in this book.

ANNOTATED BIBLIOGRAPHY

Bazarov, I. P. Thermodynamics; New York: Pergamon Press, 1964, p. 74–78.

Born, M. *Physik. Z.* 22: 218, 249, 289, 1921.

Callen, H. B. *Thermodynamics and Introductory Thermostatistics*, 2nd ed.; New York: Wiley, 1985.

- This book is particularly good on phase transitions and critical phenomena.

Carathéodory, C. *Math. Ann.* 67: 355, 1909.

Chandler, D. *Introduction to Modern Statistical Mechanics*; Oxford, U.K: Oxford University Press, 1987.

- Contains topics that are generally not found in elementary treatments.

de Heer, J. *Phenomenological Thermodynamics*; [place]: Prentice-Hall, 1986, p. 36.

Desloge, E. A. *Thermal Physics*; [place]: Holt, Rinehart and Winston, 1968.

- The approach is postulational and thorough.

Giaugque, W. E., and Archibald, R. C. *J. Am. Chem. Soc.* 59: 566, 1937.

Guggenheim, E. A. *Thermodynamics: An Advanced Treatment for Chemists and Physicists*, 5th ed.; Amsterdam: North-Holland, 1967, p. 390.

- This book is highly authoritative and highly opiniated. The author mentions in his preface that a reviewer had suggested the subtitle "Pride and Prejudice."

Landau, L. D., and Lifshitz, E. M. *Statistical Physics*; London: MIT Press, 1966.

- A rigorous and highly original book.

Thermodynamics and Introductory Statistical Mechanics, by Bruno Linder
ISBN 0-471-47459-2 © 2004 John Wiley & Sons, Inc.

Ott, J. Bevan and Boerio-Goates, J. *Chemical Thermodynamics: Advanced Applications*; New York: Academic, 2000.

Pippard, A. B. *Elements of Classical Thermodynamics*, Cambridge University Press, 1960. This book stresses problem areas (conceptual problems) and anticipates student difficulties.

Reiss, H. *Methods of Thermodynamics*; [place]: Blaisdell, 1965.

- This book is modeled after Pippard but slanted more toward chemistry.

Richards, T. W. *Z. Physk. Chem.* 42: 129, 1902.

Rushbrooke, G. S. *Introduction to Statistical Mechanics*; London: Oxford University, 1951.

- This is an easy to read elementary book on statistical mechanics.

Schilpp, P. A. (Ed.) Autobiographical Notes. Albert Einstein: Philosopher-Scientist; [place]: Open Court Publishing, 1949.

Stanley, H. E. Introduction to Phase Transitions and Critical Phenomena; Oxford, UK: Oxford University Press, 1971.

Vanderslice, J. T., Shamp, H. W., and Mason, E. A. *Thermodynamics*; [place]: Prentice-Hall, 1966.

- This book has a particularly good treatment of open systems.

Waldram, J. R. *The Theory of Thermodynamics*; [place]: Cambridge University, 1985.

- The "theory" here refers to statistical mechanical interpretation.

Wall, F. T. Mathematical definition. In: *Chemical Thermodynamics*; [place]: W. H. Freeman and Co., 1965.

APPENDIX I

HOMEWORK PROBLEM SETS

PROBLEM SET I

Chapter 3

1. One mole of an ideal gas is expanded isothermally and reversibly at $T = 300$ K from 10 to 1 atm. Calculate the work, w.

2. One mole of an ideal gas is expanded isothermally and irreversibly at 300 K in two steps: (1) by suddenly reducing the external pressure from 10 to 7 atm followed (2) by another sudden reduction of the pressure from 7 to 1 atm. Calculate w.

3. Starting with $dE = dq - PdV$, show that

(a) $dq = C_V dT + [P + (\partial E/\partial V)_T]\, dV$

(b) $\left(\frac{\partial C_V}{\partial V}\right)_T = \left[\frac{\partial}{\partial T}\left(\frac{\partial E}{\partial V}\right)_T\right]_V$

(c) dq is not an exact differential

4. Show that $dq = \left[\left(\frac{\partial E}{\partial P}\right)_T + P\left(\frac{\partial V}{\partial P}\right)_T\right] dp + \left(\frac{\partial H}{\partial T}\right)_P dT$.

5. Show that $C_P - C_V = TV\frac{\alpha^2}{\kappa}$.

Thermodynamics and Introductory Statistical Mechanics, by Bruno Linder
ISBN 0-471-47459-2 © 2004 John Wiley & Sons, Inc.

6. (a) If a gas obeys the equation of state $PV(1 - bP) = RT$, show that for an isothermal expansion the reversible work will be

$$-w = RT/(1 - bP_2) - RT(1 - bP_1) + RT \ln[P_1(1 - bP_2)/P_2(1 - bP_1)]$$
$$= P_2 V_2 - P_1 V_1 + RT \ln(P_1^2 V_1 / P_2^2 V_2)$$

(Hint) Integrate by parts to evaluate $\int_{V_1}^{V_2} P dV$

(b) If the gas is heated at constant pressure, show that $-w = \frac{RT_2}{1-b_2P} - \frac{RT_1}{1-b_1P}$

7. One mole of an ideal gas is contained in a cylinder provided with a tightly fitting piston that is not free of friction. To cause this piston to move, one must apply a constant extra force in the direction of movement. If this friction force is divided by the area of the piston, it reduces to a pressure equivalent of friction, P_f.

(a) Derive expressions for the work and heat attending the isothermal expansion of the ideal gas from a pressure P_1 to P_2. If the gas is now compressed from P_2 to P_1, what are q and w?

(b) Calculate in calories the values of q and w for 1 mol of the ideal gas expanded irreversibly from 1 atm to 0.5 atm if $P_f = 0.1$ atm and temperature $= 25°C$.

PROBLEM SET II

Chapters 3 and 4

1. Starting with $dq = dE + PdV$, show that

(a) $dq = \left(\frac{\partial E}{\partial P}\right)_V dP + \left(\frac{\partial H}{\partial V}\right)_P dV$

(b) $dq = C_V \left(\frac{\partial T}{\partial P}\right)_V dP + C_P \left(\frac{\partial T}{\partial V}\right)_P \partial V$

2. (a) Show that for an elastic hard-sphere gas, obeying the equation of state $P(V - b) = nRT$, undergoing a transition from state 1 to 2 at constant temperature

$$q = -w = nRT \ln \frac{(V_2 - b)}{(V_1 - b)}$$

(b) Show that for an elastic hard-sphere gas

$$C_P - C_V = nR$$

3. Show that for 1 mol of an elastic hard-sphere gas undergoing an adiabatic change

$$T_2/T_1 = [(V_1 - b)/(V_2 - b)]^{\gamma - 1}$$

4. Obtain an expression for the isothermal work at constant temperature associated with the transition from state 1 to state 2 of 1 mol of a gas obeying the van der Waals Equation of State,

$$\left(P + \frac{a}{V^2}\right)(V - b) = RT$$

5. One mole of an elastic hard-sphere gas with $C_V = (3/2)R$ and $b = 0.025$ liter and an initial temperature of 300 K is compressed reversibly from 10 to 1 liter. Calculate q, w, ΔE, and ΔH, if the process is carried out

   (a) Isothermally

   (b) Adiabatically

PROBLEM SET III

Chapter 4

1. Derive

$$\left(\frac{\partial C_V}{\partial V}\right)_T = T\left(\frac{\partial^2 P}{\partial T^2}\right)_V$$

2. (a) One mole of an elastic hard-sphere gas is compressed isothermally and reversibly from 600 to 300 cm^3 at 300 K, $b = 20$ cm^3. Calculate q, w, ΔE, ΔH, and ΔS.

   (b) If the same gas is compressed irreversibly at 300 K from 600 to 300 cm^3, by applying constant pressure of 2 atm, what are the values of q, w, ΔE, ΔH, and ΔS?

3. Prove that two reversible adiabats cannot intersect.

4. Show that:

$$\left(\frac{\partial S}{\partial T}\right)_P = \frac{C_P}{T};$$

$$\left(\frac{\partial S}{\partial P}\right)_T = \frac{1}{T}\left[\left(\frac{\partial H}{\partial P}\right)_T - V\right]$$

5. Derive expressions for μ_J and μ_{JT} for
   (a) Elastic hard-sphere gas
   (b) van der Waals gas

6. Derive expressions for ΔE and ΔH for
   (a) Isothermal change of an elastic hard-sphere gas
   (b) Isothermal change of a van der Waals gas
   (c) Adiabatic change of an elastic hard-sphere gas

7. One mole of supercooled water at $-5°C$ is transformed to ice at $-5°C$. Calculate ΔS. The heat of fusion of ice at the normal melting point of 273.15 K and 1 atm is 6000 J/mol; the heat capacities of supercooled water and ice are, respectively, C_P (liquid H_2O) = 75.3 $J \cdot K^{-1} \cdot mol^{-1}$ and C_P (solid H_2O) = 37.7 $J \cdot K^{-1} \cdot mol^{-1}$.

PROBLEM SET IV

Chapters 5 and 6

1. Derive the Maxwell relations.

2. Derive the Gibbs-Helmholtz equations.

3. (a) Obtain from $E = E$ $(S, V, n_1, \ldots, n_r)$ by a Legendre transformation the function $\phi = \phi(T, P, \mu_1, \ldots, \mu_r)$.
   (b) Verify that ϕ is a function of the variables $T, P, \mu_1, \ldots, \mu_r$.
   (c) Relate $\left(\frac{\partial P}{\partial T}\right)_{\mu_i}$ and $\left(\frac{\partial P}{\partial \mu_i}\right)_{T;\mu_{j \neq i}}$ to $S, V, n_1, \ldots, n_r$.

4. Show that $\left(\frac{\partial G}{\partial n_i}\right)_{T,P;n_{j \neq i}} = \left(\frac{\partial A}{\partial n_i}\right)_{T,V;n_{j \neq i}} = \left(\frac{\partial H}{\partial n_i}\right)_{S,P;n_{j \neq i}} = \left(\frac{\partial E}{\partial n_i}\right)_{S,V;n_{j \neq i}}$

5. Show that $\lim_{T \to 0} C_P = 0$

6. The molar heat capacity of SO_2 (Giaugque and Archibald, 1937) is

T (K)	C_P (cal/K mol)	T (K)	C_P (cal/K mol)
15 (solid)	0.83	110	11.97
20	1.66	120	12.40
25	2.74	130	12.83
30	3.79	140	13.31
35	4.85	150	13.82
40	5.78	160	14.33
45	6.61	170	14.85

T (K)	C_P (cal/K mol)	T (K)	C_P (cal/K mol)
50	7.36	180	15.42
55	8.02	190	16.02
60	8.62	197.64 (liquid)	20.97
70	9.57	220	20.86
80	10.32	240	20.76
90	10.93	260	20.66
100	11.49	263.1 (gas)	4.65

The heat fusion is 1,769 cal/mol at the normal melting point of 197.64 K, and the heat of vaporization is 5,960 cal/mol at the normal boiling point of 263.1 K. Calculate the entropy of gaseous SO_2 at its boiling point and 1 atm pressure.

PROBLEM SET V

Chapter 7

1. (a) A system consists of two parts ("phases") separated by a semiconducting wall that allows heat to flow only from phase "2" to phase "1." Use the enthalpy function, H, to show that the equilibrium temperatures $T^{(1)}$ and $T^{(2)}$ of the 2 phases must satisfy the relation $T^{(1)} \geq T^{(2)}$.

   (b) If the partition in *part a* is semi-permeable to chemical special "*i*" such that *i* can flow from phase "1" to phase "2." Use the Hemholtz free energy, A, to obtain a relation between the equilibrium chemical potention $\mu_i^{(1)}$ and $\mu_i^{(2)}$ of the two phases.

2. Show that, for a homogeneous system to be stable, the condition $\left(\frac{\partial P}{\partial V}\right)_T < 0$ must be met. (Use Helmholtz function, A, to prove this. Why the Helmholtz?)

PROBLEM SET VI

Chapters 7 and 8

1. Show that if $\delta^{(1)}G$ and $\delta^{(2)}G$ are zero, then $\delta^{(3)}G$ is also zero.

2. Develop conditions for stable equilibrium using the enthalpy function H. In particular, what can you say about T, P, μ_i, and their derivatives?

3. Derive explicit expression for $\overline{H}(T, P), \overline{S}(T, P)$, $\mu(T, P)$, and f in terms of
   (a) The second viral coefficient
   (b) The van der Waals equation of state constants a and b

4. The chemical potential of an ideal gas (i.e., a gas in the limit $P \rightarrow 0$) can be derived by statistical mechanics and has the form

$$-\left(\frac{\mu - H_0}{RT}\right) = \ln\frac{T^{\frac{7}{2}}}{P\Theta}$$

   where H_0 and Θ are constants.
   (a) Derive expressions for $\mu°$, $\overline{S}°$, and $\overline{H}°$ in terms of these constants and whatever variables are needed.
   (b) Write expressions for (the real gas) quantities $\mu(T, P)$, $\overline{S}(T, P)$, and $\overline{H}(T, P)$ in terms H_0, Θ, and needed variables.

PROBLEM SET VII

Chapters 9 and 10

1. At 298.15 K and 1.00 atm, ΔG for the conversion of rhombic sulfur to monoclinic sulfur is 18 cal/mol. Which of the two phases is the more stable under these conditions and why? The density of rhombic sulfur is 1.96 g/cm and that of monoclinic sulfur is 2.07 g/cm. Estimate the minimum pressure at which the other phase would be stable at 298.15 K.

2. The molar enthalpy of a gas can be written as

$$\overline{H}(T, P) = \int^T \overline{C}_{P\rightarrow 0}dT + H_0 + \int_0^P \left[\overline{V} - T\left(\frac{\partial\overline{V}}{\partial T}\right)_P\right]dP'$$

   Obtain for a gas obeying the viral equation of state $P\overline{V} = RT + B_2(T)P$:
   (a) An expression for the heat capacity $\overline{C}_P$ (T, P) in terms of $\overline{C}_{P\rightarrow 0}$ and B_2
   (b) An expression for $\overline{C}_P(T, P)$ in terms of $\overline{C}_{P\rightarrow 0}$ and the van der Waals constants a and b.

3. The specific volume of liquid water is 1 cm³/g, the specific heat of vaporization is 540 cal/g, and the vapor pressure (P^V) is 1 atm. If the liquid (or applied) pressure (P^l) is 1,000 atm, what is the boiling temperature?

4. Derive expressions for ΔG_{mix}, ΔS_{mix}, ΔH_{mix}, and ΔV_{mix} for an ideally dilute solution, using Convention II.

PROBLEM SET VIII

Chapters 11 and 12

1. If a liquid surface is increased adiabatically, will the temperature rise, fall, or remain the same?

2. Starting with $dE = TdS - PdV + Ydy + \mu dn$, derive expressions for the following:
   (a) $d\Psi$ where Ψ is a function of the natural variables S, P, y, and n
   (b) $d\phi$ where ϕ is a function of the natural variables T, P, Y, and n

3. (a) Show by thermodynamics arguments that if a rubber strip is in stable equilibrium then $\left(\frac{\partial f}{\partial L}\right)_T > 0$.
   (b) Show that the result of *part a* requires that $\phi(L)$ in the equation of state $f = T\phi(L)$ must be an increasing function of L, i.e., $\frac{d\phi}{dL} > 0$.
   (c) Determine the sign of $\alpha = \frac{1}{L}\left(\frac{\partial L}{\partial T}\right)_f$.
   (d) Show that E is independent of L.
   (e) The rubber strip is stretched under isothermal conditions. Will the heat q be positive, negative, or zero?

4. Show that the stability conditions for thermodynamic potentials (E, H, A, G) leads to

$$\left(\frac{\partial^2 G}{\partial P^2}\right)_{T_1 n_i} \leq 0 \qquad \left(\frac{\partial^2 A}{\partial T^2}\right)_{V_1 n_i} \leq 0$$

PROBLEM SET IX

Chapters 13–16

1. Atomic chlorine consists of ^{35}Cl (75%) and ^{37}Cl (25%). What fraction of molecular chlorine Cl_2 is

$$^{35}Cl - {}^{35}Cl \qquad {}^{35}Cl - {}^{37}Cl \qquad {}^{37}Cl - {}^{37}Cl$$

2. Starting with $\Omega_D^{CMB} = \prod_i \frac{g_i^{n_i}}{N_i!}$ and using $\frac{\partial \ln \Omega_D^{CMB}}{\partial N_i} = \alpha + \beta \varepsilon_i$. Derive an expression for N_i/N in terms of α, β, g_i, and ε_i.

3. A system consists of 1,000 particles, the energy levels are $\varepsilon_1 = 1$ and $\varepsilon_2 = 2$ units, the lowest level is 3-fold degenerate, and the upper level is

2-fold degenerate. The total energy is 1,200 units. What is the population of the two levels (in the most probably distribution) using Maxwell-Boltzman statistics?

4. Consider a system of N distinguishable particles partitioned among two nondegenerate energy levels of energy O and ε. Derive an expression for the energy and the temperature in terms of N_1, the number of particles in the upper state. Show that, as the energy increases past the value $\frac{1}{2} N_\varepsilon$, the temperature approaches $+\infty$, changes discontinuously to $-\infty$, and takes negative values for higher energies.

PROBLEM SET X

Chapters 13–16

1. The vibrational energy of a diatomic molecule is $E_v = \left(v + \frac{1}{2}\right) h v_o$, where v is a quantum number that runs from 0 to ∞, h is Planck's constant, and v_o is the vibration frequency. For N_2, $v_o = 6.98 \times 10^{13}$ s^{-1}. Calculate the ratio of the $v = 1$ to $v = 0$ population (i.e., N_1/N_0) at

   (a) 25°C
   (b) 800°C
   (c) 3,000°C

2. Show that for an ideal gas obeying (CMB statistics):

$$S = kN \ln \frac{q}{N} + kNT \left(\frac{\partial \ln q}{\partial T}\right)_{NVv} + kN$$

How does the "zero of energy" affect the entropy?

3. Calculate: $E - E_0$, $H - E_0$, $A - E_0$, $\mu - \varepsilon_0$, Cv of 1 mol of He gas at $T = 273$ K, and $V = 22.4$ liters. The molecular weight (MW) of He is 4. The ground state electronic degeneracy, $(g_{el,0})$ is 1.

4. Calculate the equilibrium constant K_c for the reaction $^{35}Cl_2 \rightleftharpoons 2^{35}Cl$ at $T = 2,000$ K. The vibration frequency (v_o) of $Cl_2 = 1.694 \times 10^{13}$ s^{-1}, $\theta_v = 813$ K, and $\theta_R = 0.351$ K. The molar dissociation energy (D_o) of Cl_2 is 238.9 kJ. The ground state electronic degeneracies are $g_{el,0} = 1$ for Cl_2 and $g_{el,0} = 4$ for Cl. The molecular weight of ^{35}Cl is 35.

APPENDIX II

SOLUTIONS TO PROBLEMS

SOLUTION TO SET I

1. $PV = RT; w = - \int P_{ex}dV = \int_1^2 PdV$

$$= -RT \int_1^2 dV/V = -RT \ln V_2/V_1$$

$w = +RT \ln P_2/P_1 = -RT \ln 10/1$

$\quad = -0.08206 \, atm \cdot L \cdot K^{-1} \cdot mol^{-1} \times 300 \, K \ln 10/1$

$\quad = -56.7 \, atm \cdot L \cdot mol^{-1}$

$\qquad (= -56.7 \, atm \cdot L \cdot mol^{-1} \times 8.31 \, J/0.08206 \, atm \, L$

$\qquad = -5.74 \times 10^3 \, J/mol).$

2. $P_1 = 10 \, atm: V_1 = RT/P_1$

$V_1 = 0.08206 \, atm \cdot L \cdot K^{-1} \cdot mol^{-1} \times 300 \, K/10 \, atm$

$\quad = 2.46 \, L/mol$

$P_2 = 7 \, atm: V_2 = RT/P_2 = 0.08206 \, atm \cdot L \cdot K^{-1} \cdot mol^{-1} \times 300 \, K/7 \, atm$

$\quad = 3.52 \, L/mol$

Thermodynamics and Introductory Statistical Mechanics, by Bruno Linder
ISBN 0-471-47459-2 © 2004 John Wiley & Sons, Inc.

$$P_3 = 1 \text{ atm}: V_3 = RT/P_3$$
$$= 0.08206 \text{ atm} \cdot L \cdot K^{-1} \cdot \text{mol}^{-1} \times 300 \text{ K}/1 \text{ atm}$$
$$= 24.62 \text{ L/mol}$$
$$-w = [7(3.52 - 2.46) + 1(24.62 - 3.52)] \text{ atm} \cdot L \cdot \text{mol}^{-1}$$
$$= 28.51 \text{ atm} \cdot L \cdot \text{mol}^{-1}$$

Note: If the work were carried out reversibly, from 10 to 1 atm, it would be $-56.7 \text{ atm} \cdot L \cdot \text{mol}^{-1}$, the same as in *Problem 1*.

3. $dE = dq - PdV$

(a) $dq = dE + PdV = (\partial E/\partial T)_V dT + [(\partial E/\partial V)_T + P]dV$
$$= C_V dT + [P + \partial E/\partial V]_T dV$$

(b) $dE = (\partial E/\partial T)_V + (\partial E/\partial V)_T dV = C_V dT + (\partial E/\partial V)_T dV$
Because dE is an exact differential, the reciprocity relation gives

$$(\partial C_V/\partial V)_T = [\partial/\partial T(\partial E/\partial V)_T]_V$$

(c) If dq were an exact differential, then by *solution 3a* $(\partial C_V/\partial V)_T$ would have to be equal to $[\partial/\partial T(P + (\partial E/\partial V)_T)]_V$ but it is not according to *solution 3b*. Hence, dq is not exact.

4. $dq = dH - V \, dP;$ $H = E + PV$
$dq = (\partial H/\partial P)_T + (\partial H/\partial T)_P - V \, dP$
$$= (\partial E/\partial P)_T dP + P(\partial V/\partial P)_T dP + V \, dP + (\partial H/\partial T)_P dT - V \, dP$$
$$= [(\partial E/\partial P)_T + P(\partial V/\partial P)_T]dP + (\partial H/\partial T)_P$$

5. *Equation 3-45* gives,

$$C_P - C_V = T(\partial P/\partial T)_V(\partial V/\partial T)_P$$
$$= -TV[(1/V)(\partial V/\partial T)_P]/[1/V(\partial V/\partial P)_T] \times [1/V(\partial V/\partial T)_P]$$
$$= TV\alpha^2/\kappa$$

6a. $-w = \int PdV = P_2 V_2 - P_1 V_1 - RT \int dP/P(1 - bP);$

$V = RT/(1 - bP)P$

$-w = P_2 V_2 - P_1 V_1 - RT \int [1/P + b/(1 - bP)]dP$

$$= P_2 V_2 - P_1 V_1 - RT\{\ln P_2/P_1 - \ln[(1 - bP_2)/(1 - bP_1)]\}$$
$$= RT/(1-bP_2) - RT/(1 - bP_1) + RT \ln[P_1(1- bP_2)/P_2(1- bP_1)]$$

From $V = RT/[P(1 - bp)] \Rightarrow 1/(1 - bp) = PV/RT$. Hence,

$$1/(1 - bp_1) = P_1V_1/RT; \qquad 1/(1 - bp_2) = P_2V_2/RT$$

Thus, $-w = P_2V_2 - P_1V_1 + RT \ln\{P_1/P_2[(P_1V_1/RT)/(P_2V_2/RT)]\}$
$$= P_2V_2 - P_1V_1 + RT \ln(P_1^2V_1/P_2^2V_2)$$

6b. $-w = PV_2 - P_1V_1; \qquad P$ constant $\Rightarrow P = P_2 = P_1$

In general, b varies with T. Let us assume that at $T_2 \quad b = b_2$ and at $T_1 \quad b = b_1$

Then, $V_2 = RT_2/[P(1 - b_2P)]; \qquad V_1 = RT_1/[P(1 - b_1P)]$
and $-w = RT_2/(1 - b_2P) - RT_1/(1 - b_1P)$

7a. Compression: $P_{ex,c} = P + P_f; \qquad$ Expansion: $P_{ex,e} = P - P_f$

$$\text{Expansion: } q = -w = \int_1^2 P_{ex,e}dV = \int_1^2 dV(RT/V - P_f)$$
$$= RT \ln V_2/V_1 - P_f(V_2 - V_1)$$
$$\text{Compression: } q = -w = \int_2^1 P_{ex,c}dV = \int_2^1 dV(RT/V + P_f)$$
$$= -RT \ln V_2/V_1 - P_f(V_2 - V_1)$$
$$\text{Overall: } q = -w = -2P_f(V_2 - V_1)$$

Because $V_2 > V_1$, the overall q is negative and w is positive. Thus, by our convention, work associated with frictional forces is converted into heat.

7b.
$q = -w = RT \ln P_1/P_2 - RT(1/P_2 - 1/P_1)P_f$
$= 1.987 \, \text{cal} \cdot \text{K}^{-1} \cdot \text{mol}^{-1} \times 298 \, \text{K}[\ln 1/0.5 - 0.1(1/0.5 - 1)]$
$= 351 \, \text{cal/mol}$

SOLUTION TO SET II

1a. $dq = dE - dw = (\partial E/\partial P)_v dP + [(\partial E/\partial V)_P + P]dV$
$$= (\partial E/\partial P)_V dP + (\partial H/\partial V)_P dV$$
1b. $dq = (\partial E/\partial T)_V(\partial T/\partial P)_V dP + (\partial H/\partial T)_P(\partial T/\partial V)_P dV$
$$= C_v(\partial T/\partial P)_V dP + C_P(\partial T/\partial V)_P dV$$

2a. Applying the Thermodynamic Equation of State

$$(\partial E/\partial V)_T = T(\partial P/\partial T)_V - P$$

to the equation of state $P = nRT/(V - b)$ gives $nRT/(V - b) - P = 0$
  Thus, at constant T, the internal energy E is independent of V, similar to an ideal gas, and $\Delta E = 0$. Accordingly,

$$q = -w = nRT \int_1^2 dV/(V - b) = nRT \ln[(V_2 - b)/(V_1 - b)]$$

2b. $C_P - C_V = (\partial H/\partial T)_P - (\partial E/\partial T)_V$
$$= (\partial E/\partial T)_P + P(\partial V/\partial T)_P - (\partial E/\partial T)_V$$
$(\partial E/\partial T)_P = (\partial E/\partial T)_V + (\partial E/\partial V)_T(\partial E/\partial T)_P = (\partial E/\partial T)_V + 0$
$C_P - C_V = P(\partial V/\partial T)_P; \quad V = (nRT/P) + b; \quad (\partial V/\partial T)_P = nR/P$
Thus, $C_P - C_V = nR$.

3. $dE = dq + dw; \qquad dE + PdV = 0(\text{adiabatic})$

$$C_V dT + RTdV/(V - b) = 0$$

$$C_V dT/T + RdV/(V - b) = 0$$

$$C_V \ln(T_2/T_1) + R \ln[(V_2 - b)/V_1 - b)] = 0 \qquad C_V \text{ constant}$$

$$R = C_P - C_V$$

$$\ln\{(T_2/T_1)[(V_2 - b)/(V_1 - b)]^{(\gamma-1)}\} = 0$$

$$T_2/T_1 = [(V_1 - b)/(V_2 - b)]^{(\gamma-1)}$$

4. $-w = \int_1^2 P \, dV; \qquad P = RT/(V - b) - a/V^2$

$$-w = RT \int_1^2 [dV/(V - b) - a \, dV/V^2]$$

$$= RT \ln[(V_2 - b)/(V_1 - b)] + a(1/V_2 - 1/V_1)$$

5a. $q = -w = \int_1^2 P dV = RT \int_1^2 dV/(V - b) = RT \ln[(V_2 - b)/V_1 - b)]$

$R = 8.314 \, J \cdot K^{-1} \cdot mol^{-1}; V_2 = 1 \, L; V_1 = 10 \, L$

$q = -w = 8.314 \, J \cdot K^{-1} \cdot mol^{-1} \times 300 \, K \ln[(1 - 0.025)/(10 - 0.025)]$

$$= -5.80 \, kJ/mol$$

$$\Delta E = 0 (\text{isothermal}); \qquad \Delta H = \Delta E + \Delta(PV) = P_2 V_2 - P_1 V_1$$

$$\Delta H = 8.314\,\text{J}\cdot\text{K}^{-1}\cdot\text{mol}^{-1} \times 300\,\text{K}[1/(1-0.025) - 10/(10-0.025)]$$

$$= 57.70\,\text{J/mol}$$

5b. $q = 0; w = \Delta E = \displaystyle\int_1^2 C_V dT = (3/2)\,R(T_2 - T_1)$

$$T_2 = T_1[(V_1 - b)/(V_2 - b)]^{(\gamma-1)} = 300\,\text{K}[(10-0.025)/(1-0.025)]^{2/3}$$

$$= 1{,}413.8\,\text{K}$$

$$\Delta E = 13.89\,\text{kJ/mol}; \Delta H = \Delta E + RT_2 V_2/(V_2 - b) - RT_1 V_1/(V_1 - b)$$

$$= 23.45\,\text{kJ/mol}$$

SOLUTION TO SET III

1. $dS = dq_{rev}/T = dE/T + PdV/T = 1/T[(\partial E/\partial V)_T dV + (\partial E/\partial T)_V dT] + PdV/T$

Applying the thermodynamic equation of state, $(\partial E/\partial V)_T = T(\partial P/\partial T)_V - P$, gives

$$dS = (\partial P/\partial T)_V dV + (\partial E/\partial T)_V dT$$
$$= (\partial P/\partial T)_V dV + (C_V/T)dT$$

Since dS is an exact differential, the reciprocity relation gives

$$(\partial^2 P/\partial T^2)_V = 1/T(\partial C_V/\partial V)_T \quad \text{or} \quad (\partial C_V/\partial V)_T = T(\partial^2 P/\partial T^2)_V$$

2a. Elastic hard-sphere gas: $P(V - b) = RT; V_2 = 300\,\text{cm}^3; V_1 = 600\,\text{cm}^3;$
  $b = 20\,\text{cm}^3$

$$w = -\int PdV = -RT\int dV/(V-b)$$
$$= -RT\ln[(300-20)/(600-20)]$$
$$= 8.314\,\text{J}\cdot\text{K}^{-1}\cdot\text{mol}^{-1} \times 300\,\text{K}\ln(580/280)$$
$$= 1.816\,\text{kJ/mol}$$
$$dE = (\partial E/\partial V)_T dV + (\partial E/\partial T)_V dT$$
$$(\partial E/\partial V)_T = T(\partial P/\partial T)_V - P = TR/(V-b) - P = 0$$

Thus, $dE = (\partial E/\partial T)_v dT = C_V dT; \Delta E = 0$ at constant T

$q = -w = -1.816\,\text{kJ/mol}$

$\Delta S = q_{rev}/T = -(1.816\,\text{kJ/mol})/300\,\text{K} = -6.053\,\text{J}\cdot\text{K}^{-1}\cdot\text{mol}^{-1}$

$$\Delta H = \Delta E + \Delta(PV) = 0 + RT[V_2/(V_2 - b) - V_1/(V_1 - b)]$$
$$= 8.314 \, J \cdot K^{-1} \cdot mol^{-1} \times 300 \, K[300/280 - 600/580]$$
$$= 92.15 \, J/mol$$

2b. The initial and final states are the same as in *solution 2a*. Consequently, the values of ΔE, ΔH and ΔS are the same because E, H, and S are state variables and independent of path. As in *solution 2a*, q and w are different:

$$q = -w = -P_{ex}(V_2 - V_1) = -2 \, atm \times (300 - 600) \, cm^3/mol$$
$$= 600 \, atm \cdot cm^3 \cdot mol^{-1}$$
$$= 600 \, atm \cdot cm^3 \cdot mol^{-1} \times 8.314 \, J/(82.06 \, atm \cdot cm^3)$$
$$= 60.79 \, J/mol$$

3. Suppose curves 1–2 and 2–3 are intersecting (reversible) adiabats. Curve 3–1 represents an isotherm. Consider the cyclic process: $1 \rightarrow 2$, $2 \rightarrow 3$, $3 \rightarrow 1$. In this process, $q_{1 \rightarrow 2} = 0$ (adiabatic transition), $q_{2 \rightarrow 3} = 0$ (adiabatic transition), and $q_{3 \rightarrow 1} > 0$ (isothermal expansion) (see Figure S3.1). $\Delta E = 0$ (cyclic process). The net result is the overall $q > 0$ and

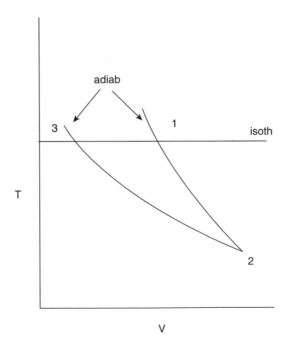

Figure S3.1   Plot of T vs. V

consequently w < 0. In other words, heat is converted entirely into work, in violation of the Kelvin-Planck Principle. The assumption that two adiabats intersect is invalid.

4.
$$dq = dH - VdP = (\partial H/\partial T)_P dT + (\partial H/\partial P)_T dP - VdP;$$
$$TdS = dq_{rev}$$
$$(\partial S/\partial T)_P = dq_{rev}/T = 1/T(\partial H/\partial T)_P = C_P/T$$
$$T(\partial S/\partial P)_T = [(\partial H/\partial P)_T - V]$$

5a. Elastic hard sphere: $P(V - b) = RT$

$$\mu_J = (\partial T/\partial V)_E = -(\partial E/\partial V)_T/(\partial E/\partial T)_V = -1/C_V(\partial E/\partial V)_T$$
$$(\partial E/\partial V)_T = T(\partial P/\partial T)_V - P = TR/(V - b) - RT/(V - b) = 0$$

Thus, $\mu_J = 0$

$$\mu_{JT} = (\partial T/\partial P)_H = -(\partial H/\partial P)_T/(\partial H/\partial T)_P$$
$$= -(1/C_P)(\partial H/\partial P)_T; V = RT/P + b$$
$$(\partial H/\partial P)_T = V - T(\partial V/\partial T)_P = RT/P + b - TR/P = b$$
$$\mu_{JT} = -b/C_P$$

5b. van der Waals Equation: $(P + a/V^2)(V - b) = RT \rightarrow P = RT/(V - b) - a/V^2$

$$(\partial E/\partial V)_T = TR/(V - b) - [RT/(V - b) - a/V^2] = a/V^2$$
$$\mu_J = -(1/C_V)a/V^2$$

To determine $(\partial H/\partial P)_T$, one needs $(\partial V/\partial T)_P$, which is messy because in the van der Waals Equation, V is present in the denominator. It is easier to differentiate the form $PV - Pb + a/V - ab/V^2 = RT$

$$P(\partial V/\partial T)_P - (a/V^2)(\partial V/\partial T)_P + (2ab/V^3)(\partial V/\partial T)_P = R$$
$$T(\partial V/\partial T)_P - V = RT/(P - a/V^2 + 2ab/V^3) - V$$

Replacing P by $RT/(V - b) - a/V^2$ gives

$$\mu_{JT} = 1/C_P\{RT/[RT/(V - b) - 2a/V^2 + 2ab/V^3] - V\}$$

(Later, after introducing the virial form of the van der Waals equation of state, we shall see that there is a much simpler, although less exact, form of the Joule-Thomson coefficient of a van der Waals gas.)

6a. $\Delta E = \int (\partial E/\partial V)_T dV + \int (\partial E/\partial T)_V$

$\qquad = 0 + 0 \qquad$ for an elastic hard sphere.

$\Delta H = \int (\partial H/\partial P)_T dP + \int (H/\partial T)_P dT$

$\qquad = \int_1^2 b dP = b(P_2 - P_1)$

(see *problem* 5 in corresponding solution set)

6b. $\Delta E = \int_1^2 (a/V^2) dV = -a(1/V_2 - 1/V_1)$

$\Delta H = \Delta E + \Delta(PV) = -a(1/V_2 - 1/V_1) + P_2 V_2 - P_1 V_1$

$P_2 = RT/(V_2 - b) - a/V_2^2; \quad P_1 = RT(V_1 - b) - a/V_1^2$

$\Delta H = -2a(1/V_2 - 1/V_1) + RT[V_2/(V_2 - b) - V_1/(V_1 - b)]$

6c. $\Delta E(= w) = \int_1^2 C_V dT = C_V(T_2 - T_1)$ for elastic hard sphere, if C_V is constant.

$$\Delta H = \int (\partial H/\partial T)_P dT + \int (\partial H/\partial P)_T dT$$

$$= \int_1^2 C_P dT + b \int_1^2 dP$$

$$= C_P(T_2 - T_1) + b(P_2 - P_1)$$

for elastic hard sphere if C_P is constant.

7. This is an irreversible process. One must choose a reversible path.

$$H_2O(\text{liquid}, 273.15\,K) \qquad (2) \rightarrow_{rev} \qquad H_2O(\text{solid}, 273.15\,K)$$
$$\uparrow^{rev} (1) \qquad\qquad\qquad \downarrow_{rev} (3)$$
$$H_2O(\text{liquid}, 268.15\,K) \rightarrow_{irr} \qquad H_2O(\text{solid}, 268.15)$$

$\Delta S = \Delta S(1) + \Delta S(2) + \Delta S(3)$

$\qquad = [75.3 \ln(273.15/268.15)$

$\qquad\quad + (-6000/273.15) + 37.7 \ln(263.15/273.15)]J \cdot K^{-1} \cdot mol$

$\qquad = [(75.3 - 37.7) \ln(273.15/268.15) - 6000/273.15]J \cdot K^{-1} \cdot mol^{-1}$

$\qquad = -20.56\,J \cdot K^{-1} \cdot mol^{-1}$

SOLUTION TO SET IV

1. $dE = TdS - PdV;$ reciprocity $\rightarrow$ $(\partial T/\partial V)_S = -(\partial P/\partial S)_V$
$dH = TdS + VdP$ $(\partial T/\partial P)_S = (\partial V/\partial S)_P$
$dA = -SdT - PdV$ $(\partial S/\partial V)_T = (\partial P/\partial T)_V$
$dG = -SdT + VdP$ $(\partial S/\partial P)_T = (\partial V/\partial T)_P$

2. $dA = -SdT - PdV; A = E - TS; A/T = E/T - S$

$$[\partial(A/T)/\partial T]_V = 1/T(\partial A/\partial T)_V - A/T^2 = -S/T - E/T^2 + S/T$$
$$= -E/T^2$$

or

$$[\partial(A/T)/\partial(1/T)]_V = -T^2[(\partial A/T)/\partial T]_V = E$$
$$dG = -SdT + VdP; G = H - TS; G/T = H/T - S$$
$$[\partial(G/T)/\partial T]_P = 1/T[\partial(G/T)/\partial T]_P - G/T^2$$
$$= -S/T - H/T^2 + S/T = -H/T^2$$

or

$$[\partial G/T)/\partial(1/T)]_P = -T^2[\partial(G/T)/\partial T]_P = H$$

3a. $\phi = E - S(\partial E/\partial S)_{V,\underline{n_i}} - V(\partial E/\partial V)_{S,\underline{n_i}} - \Sigma_i n_i(\partial E/\partial n_i)_{S,V;n_{j \neq i}}$

$$= TS - PV + \Sigma_i \mu_i n_i - ST + VP - \Sigma_i \mu_i n_i = 0$$

3b.

$$d\phi = dE - TdS - SdT + PdV + VdP - \Sigma_i n_i d\mu_i - \Sigma_i \mu_i dn_i$$
$$dE = TdS - PdV + \Sigma_i \mu_i dn_i$$
$$\rightarrow d\phi = -SdT + VdP - \Sigma_i n_i d\mu_i = 0$$

3c. $(\partial P/\partial T)_{\underline{\mu_i}} = S/V$

$$V(\partial P/\partial \mu_i)_{T,\mu_{j \neq i}} - n_i = 0 \rightarrow (\partial P/\partial \mu_i)_{T,\mu_{j \neq i}} = n_i/V$$

4. *Equation 5-28c* gives derivation of

$$(\partial G/\partial n_i)_{T,P; n_{j \neq i}} = (\partial E/\partial n_i)_{S,V; n_{j \neq i}}$$

We derive here the relation between G and H.

$$dG = -SdT + VdP + \Sigma_i(\partial G/\partial n_i)_{T,P;\,n_{j\neq i}}\,dn_i$$
$$dH = TdS + VdP + \Sigma_i(\partial H/\partial n_i)_{S,P;\,n_{j\neq i}}\,dn_i$$
$$G = H - TS; \rightarrow dG = dH - TdS - SdT$$

Thus,

$$-SdT + VdP + \Sigma_i(\partial G/\partial n_i)_{T,P;\,n_{j\neq i}}\,dn_i$$
$$= VdP - SdT + \Sigma_i(\partial H/\partial n_i)_{S,P;\,n_{j\neq i}}\,dn_i$$

Thus,

$$(\partial G/\partial n_i)_{T,P;n_{j\neq i}} = (\partial H/\partial n_i)_{S,P;n_{j\neq i}}$$

5. $G = TS; \quad (\partial G/\partial T)_P = -S = (G - H)/T$

$$\lim_{T\to 0}(\partial G/\partial T)_P = \lim_{T\to 0}(G - H)/T = 0/0$$

Applying l'Hospital's Rule gives

$$\lim_{T\to 0}(\partial G/\partial T)_P = \lim_{T\to 0}[(\partial G/\partial T)_P/(\partial T/\partial T) - (\partial H/\partial T)_P/(\partial T/\partial T)]$$
$$= [0/1 - \lim_{T\to 0} C_P/1] = 0$$
$$\rightarrow \lim_{T\to 0} C_P = 0$$

6. Plot C_P/T vs. T (see Fig S4.1). There are discontinuities at $T =$ 197.64 K (solid-liquid transition) and at $T = 263.1$ K (liquid-gas transition). No data are given for $T < 15$ K. Using the Debye Cube Law,

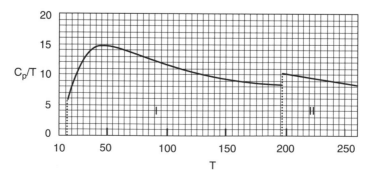

Figure S4.1 Plot of C_p vs. T

$C_P = aT^3$, for temperatures between 0 and 15 K yields $S(15\,K) - S(0\,K) = S_{15\,K} = \int_0^{15}(C_P/T)dT = (1/3)aT^3$.

The value of C_P at $T = 15\,K$ is $0.83 \cdot cal \cdot K^{-1} \cdot mol^{-1}$ producing

$$a = 0.83/15^3\, cal \cdot K^{-4} \cdot mol^{-1} = 2.46 \times 10^{-4}\, K^{-4} \cdot mol^{-1}$$

T	C_P/T	T	C_P/T
15	5.53×10^{-2}	110	1.09×10^{-1}
20	8.53×10^{-2}	120	1.03×10^{-1}
25	1.096×10^{-1}	130	9.87×10^{-2}
30	1.26×10^{-1}	140	9.51×10^{-2}
35	1.386×10^{-1}	150	9.21×10^{-3}
40	1.45×10^{-1}	160	8.46×10^{-2}
45	1.47×10^{-1}	170	8.74×10^{-2}
50	1.47×10^{-1}	180	8.57×10^{-2}
55	1.46×10^{-1}	190	8.43×10^{-2}
60	1.44×10^{-1}	197.64	1.06×10^{-1}
70	1.37×10^{-1}	220	9.48×10^{-2}
80	1.29×10^{-1}	240	8.65×10^{-2}
90	1.21×10^{-1}	260	7.95×10^{-2}
100	1.51×10^{-1}	263.1	1.77×10^{-2}

The entropy at the boiling point $T = 363.1\,K$ is obtained by adding the following:

$$(1/3 \times 0.83 + area\ I + 1,769/197.64 + area\ II + 5,960/263.1)cal \cdot K^{-1} \cdot mol^{-1}.$$

SOLUTION TO SET V

1a. Criteria for stable equilibrium requires that a virtual variation of H, under constraints of constant S, P, and n_i, $ДH_{S,P;n_i} \geq 0$. This condition can be derived from $dE \leq TdS - PdV + \Sigma_i\mu_idn_i$. Adding $d(PV)$ to both sides of the equaton produces

$$d(E + PV) \leq TdS - PdV + \Sigma_i\mu_idn_i + PdV + VdP$$

$$dH \leq TdS + VdP + \Sigma_i\mu_idn_i$$

$$dH_{S,P;\,n_i} \leq 0 \Rightarrow ДH_{S,P;\,n_i} \geq 0$$

For stable equilibrium of systems that are "normal" (where there are no restrictions to the flow of heat, volume changes, and flow of matter), first- and second-order variations are, respectively, $\delta^{(1)}H_{S,P;n_i} = 0$ and $\delta^{(2)}H_{S,P;n_i} > 0$.

In the present problem, heat is restricted to flow only from "2" to "1," and so the first-order variation is not necessarily zero but may be positive., i.e. $\delta^{(1)}H_{S,P;\underline{n}_i} \geq 0$. Holding n_i constant (variation in P is not allowed) and varying parts of S, keeping total S fixed so that $\delta S^{(1)} = -\delta S^{(2)}$ with $\delta S^{(1)} > 0$, gives

$$\delta^{(1)}H_{S,P;\,\underline{n}_i} = T^{(1)}\delta S^{(1)} + T^{(2)}\delta S^{(2)} \geq 0$$
$$\rightarrow [T^{(1)} - T^{(2)}]\delta S^{(1)} \geq 0 \text{ and thus } T^{(1)} \geq T^{(2)}$$

1b. Using the conditions $Д A_{T,V;\,\underline{n}_i} \geq 0$ and $\delta^{(1)}A_{T,V;\,\underline{n}_i} \geq 0$ and holding all variations constant except $\delta^{(1)}_{n_i} = -\delta n_i^{(2)}$ with $\delta n_i^{(2)} > 0$, we obtain

$$\delta^{(1)}A_{T,V;\,n_i} = \mu_i^{(1)}\delta n_i^{(1)} + \mu_i^{(2)}\delta n_i^{(2)} = [\mu_i^{(2)} - \mu_i^{(1)}]\delta n_i^{(2)} \geq 0$$

or

$$\mu_i^{(2)} \geq \mu_i^{(1)}$$

2. A homogeneous system is "normal" in the sense that the partition between the two parts is fully heat conducting, fully deformable, and fully permeable. Consequently, temperature, pressure, and chemical potential of each species is uniform throughout. Thus, the first-order variation, $\delta^{(1)}A_{T,V;\,\underline{n}_i}$, is zero and we must go to the second-order variation, $\delta^{(2)}A_{T,V;\,\underline{n}_i}$, which for stable equilibrium must be positive.

Holding constant variations in n_i (T is already constant) and allowing variations in V, namely, $\delta V^{(1)} = -\delta V^{(2)}$, we get

$$\delta^{(2)}A_{T,V,n_i} = [\partial^2 A/\partial(V^{(1)})^2]_{T,V^{(2)},\underline{n}_i}[\delta V^{(1)}]^2 + [\partial^2 A/\partial(V^{(2)})^2]_{T,V^{(1)},\underline{n}_i}[\delta V^{(2)}]^2 > 0$$
$$= (-\partial P^{(1)}/\partial V^{(1)})_{T,V^{(2)},\underline{n}_i}(\delta V^{(1)})^2 + (-\partial P^{(2)}/\partial V^{(2)})_{T,V^{(1)},\underline{n}_i}(\delta V^{(2)})^2 > 0$$

Because $(-\delta V^{(1)})^2 = (-\delta V^{(2)})^2$ and the pressure is uniform, i.e., $P^{(1)} = P^{(2)} = P$ we get

$$[-(\partial P/\partial V^{(1)})_{T,V^{(2)},\underline{n}_i} - (\partial P/\partial V^{(2)})_{T,V^{(1)},\underline{n}_i}] > 0 \qquad \text{or}$$
$$[(\partial P/\partial V^{(1)})_{T,V^{(2)},\underline{n}_i} + (\partial P/\partial V^{(2)})_{T,V^{(1)},\underline{n}_i} < 0$$

If we denote the partial molar volume of species i by the symbol $\bar{V}_i$, we can write for the volume element $V^{(1)} = \Sigma_i n_i^{(1)}\bar{V}_i$ and $V^{(2)} = \Sigma_i n_i^{(2)}\bar{V}_i$. It is obvious that, since the n_i are always positive, we must have $(\partial P/\partial V)_T < 0$ as a requirement for stable equilibrium.

SOLUTION TO SET VI

1. Knowing that $\delta^{(1)}G_{T,P;n_i} = 0$ means that fluctuations can proceed in both directions, i.e., the system is "normal." Since $n_i^{(2)} + n_i^{(2)}$ is constant, we have $\delta n_i^{(1)} = -\delta n_i^{(2)}$. We also know that $\delta^{(2)}G_{T,P;n_i} = 0$, so we must go to $\delta^{(3)}G_{T,P;n_i} \geq 0$.

$$\delta^{(3)}G_{T,P;n_i} = 1/6\{[\partial^3 G^{(!)}/\partial(n_i^{(2)})^3]_{T,P;n_{j\neq i}}(\delta n_1^{(!)})^3$$
$$+ [\partial^3 G^{(2)}/\partial(n_i^{(2)})^3]_{T,P;n_{j\neq i}}(\delta n_i^{(2)})^3\} \geq 0.$$

If $\delta n_i^{(1)} > 0$, so is $(\delta n_i^{(1)})^3$ and $(\delta n_i^{(2)})^3 < 0$. Therefore, we can write

$$\delta^{(3)}G_{T,P;n_i} = 1/6\{[\partial^3 G^{(!)}/\partial(n_i^{(1)})^3]_{T,P;n_{j\neq i}}$$
$$- [\partial^3 G^{(2)}/\partial(n_i^{(2)})^3]_{T,P;n_{j\neq i}}\}(\delta n_i^{(1)})^3 \geq 0.$$

If $\delta n_i^{(1)} < 0$, so is $(\delta n_i^{(1)})^3$ and $(\delta n_i^{(2)})^3 > 0$. We have then

$$\delta^{(3)}G_{T,P;n_i} = 1/6\{[\partial^3 G^{(!)}/\partial(n_i^{(1)})^3]_{T,P;n_{j\neq i}}$$
$$- [\partial^3 G^{(2)}/\partial(n_i^{(2)})^3]_{T,P;n_{j\neq i}}\}(\delta n_i^{(1)})^3 \leq 0.$$

The only condition that is consistent is the equal sign, or $\delta^{(3)}G_{T,P;n_{j\neq i}} = 0$.

2. In $ДH_{S,P;n_i} = \delta^{(1)}H_{S,P;n_i} + \delta^{(2)}H_{S,P;n_i} + \ldots > 0$, P is fixed, but there can be variations in S and n_i.

   (a) Variation in S gives

   $$\delta^{(1)}H_{S,P;n_i} \Rightarrow [T^{(1)} - T^{(2)}] = 0 \quad \text{or} \quad T^{(1)} - T^{(2)} = T$$
   $$\delta^{(2)}H_{S,P;n_i} \Rightarrow \frac{1}{2}[T/C_P^{(1)} + T/C_P^{(2)}] > 0$$

   or

   $$C_P > 0 \text{ except when } T = 0$$

   (b) Variation in n_i gives

   $$\delta^{(1)}H_{S,P;n_{j\neq i}} \to [\mu_i^{(1)} - \mu_i^{(2)}] = 0 \quad \text{or} \quad \mu_i^{(1)} = \mu_i^{(2)} = \mu_i$$
   $$\delta^{(2)}H_{S,P;n_{j\neq i}} \to \frac{1}{2}\{[\partial\mu_i/\partial n_i^{(1)}]_{S,P;n_{j\neq i}} + [\partial\mu/\partial n_i^{(2)}]_{S,P;n_{j\neq i}}\}(\delta n_i^{(1)})^2 > 0$$

   or

   $$(\partial\mu_i/\delta n_i)_{S,P;n_{j\neq i}} > 0$$

3. In this problem, all the macroscopic variables V, H, S refer to molar quantities.

$$H(T, P) = H^0(T) + \int_0^P [V - T(\partial V/\partial T)_P dP']; \quad V = RT/P + B_2(T)$$

$$(\partial V/\partial T)_P = R/P + dB_2/dT; \quad V = T(\partial V/\partial T)_P$$

$$= (RT/P) + B_2(T) - \frac{TR}{P} - TdB_2/dT$$

(a) $H(T, P) = H^0(T) + \int_0^P [B_2(T) - TdB_2/dT] dP'$

$$= H^0 + (B_2 - TdB_2/dT)P$$

(b) $\qquad B_2(T) = b - a/RT; \quad dB_2/dT = a/RT^2$

$$\rightarrow H(T, P) = H^0(T) + (b - a/RT - Ta/RT^2)$$

$$= H^0(T) + (b - 2a/RT)P$$

$$S(T, P) = S^0(T) - R\ln(P/P^0) + \int_0^P [R/P - (\partial V/\partial T)_P] dP'$$

(a) $S(T, P) = S^0 - R\ln(P/P^0) + \int_0^P [R/P - R/P - dB_2/dT] dP'$

$$= S^0(T) - R\ln(P/P^0) - dB_2/dT$$

(b) $\quad S(T, P) = S^0(T) - R\ln(P/P^0) - aP/RT^2$

$$\mu(T, P) = \mu^0(T) + RT\ln(P/P^0) + \int_0^P [\overline{V} - RT/P'] dP'$$

(a) $\mu(T, P) = \mu^0(T) + RT\ln(P/P^0) + \int_0^P (RT/P' + B_2 - RT/P'] dP'$

$$= \mu^0(T) + RT\ln(P/P^0) + B_2(T)P$$

(b) $\quad \mu(T, P) = \mu^0(T) + RT\ln(P/P^0) + (b - a/RT)P$

$$f = P\exp[1/RT \int_0^P (\overline{V} - RT/P')dP']$$

a) $f = P\exp[B_2(T)P/RT]$

b) $f = P\exp\{(1/RT)(b - a/RT)P\}$

4a. From $-(\mu - H_0)/RT = \ln(T^{7/2}/P\Theta)$, it must be assumed that $T^{7/2}/P\Theta$ is dimensionless. We split this quantity into terms involving T, P, and Θ; however, in doing so we introduce unit quantitities T^0, P^0 to make the logarithmic terms dimensionless. We replace the logarithmic term by $[(T/T^0)^{7/2}/(P/P^0)] \times [(T^0)^{7/2}/(P^0\Theta)]$. This gives

$$\mu = H_0 - (7/2)\ln T/T^0 + RT\ln(P/P^0) + RT\ln[P^0\Theta/(T^0)^{7/2}]$$

(a) The problem states that the above expression is good only for an ideal gas. But we know from thermodynamics that for an ideal gas

$$\mu(T,P) = \mu^0(T) + RT\ln(P/P^0)$$

We therefore identify

$$\mu^0(T,P) = H_0 - (7/2)RT\ln T/T^0 + RT\ln[P^0\Theta/(T^0)^{7/2}]$$
$$S^0(T) = -(\partial\mu^0/\partial T)_P = (7/2)R\ln(T/T^0) + (7/2)R - R\ln[P^0\Theta/(T^0)^{7/2}]$$
$$H^0(T) = \mu^0(T) + TS^0(T) = H_0 + (7/2)RT$$

4b. Using the above expressions for μ^0, $\overline{H}^0$, and $\overline{S}^0$, we obtain for the real gas

$$\mu(T,P) = H_0 - (7/2)RT\ln T/T^0 + RT\ln(P/P^0) + RT\ln[P^0\Theta/(T^0)^{7/2}]$$
$$+ \int_0^P (V - RT/P')dP'$$
$$S(T,P) = (7/2)R\ln T/T^0 + (7/2)R - R\ln[P^0\Theta/(T^0)^{7/2}]$$
$$+ \int_0^P [R/P' - (\partial V/\partial T)_{P'}]dP'$$
$$H(,P) = H_0 + (7/2)RT + \int_0^P [\overline{V} - T(\partial V/\partial T)_{P'}\,dP']$$

SOLUTION TO SET VII

Note: In this set the quantities G, H, S, and V represent molar quantities.

1. At 298.15 K, rhombic sulfur, S_r, has a lower free energy per mole or chemical potential, μ_r, than monoclinic sulfur, S_m; therefore, in the reaction $S_r \rightarrow S_m$ at $T = 298.15\,K$ and $P^0 = 1$ atm, S_r is more stable.

Let $\Delta\mu = \mu_m - \mu_r$ and then $(\partial\Delta\mu/\partial P)_T = V_r - V_m = \Delta V$. The volume difference (ΔV) is as follows:

$$\Delta V = 32(1/2.07 - 1.1.96)\text{cm}^3\text{mol}^{-1} = -2.711 \times 10^{-2}\text{cm}^3/\text{mol}$$

At the equilibrium pressure, P_{eq}, S_m becomes more stable. There are several ways to calculate P_{eq}. We discuss three.

(1) $\int d\Delta\mu = \int \Delta V dP \approx \Delta V \int_1^{P_{eq}} dP \approx \Delta V(P_{eq} - 1)$

$$\Delta\mu(P_{eq}) - \Delta\mu(1 \text{ atm}) \approx \Delta V(P_{eq} - 1)$$

$$0 - (18/32) \text{ cal/mol} = -2.711 \times 10^{-2} \text{cm}^3 \cdot \text{mol}^{-1}(P_{eq} - 1)$$

$$\times 1.987 \text{ cal}/(82.06 \text{ atm} \cdot \text{cm}^3)$$

$$= -[6.57 \times 10^{-4} \text{cal}/(\text{mol atm})] \times (P_{eq} - 1)$$

$$P_{eq} = 857.8 \text{ atm}$$

(2) We can also use the expressions for condensed systems:

$$\mu(T, P) = \mu^0(T, P = 0) + PV(T, 0)[1 - \kappa P/2]$$

The quantity within square brackets is approximately one.

$$\mu_m(T, P) - \mu_m(T, P = 1 \text{ atm}) \approx (P_{eq} - 1)V_m(T, 0)$$
$$\mu_r(T, P) - \mu_r(T, P = 1 \text{ atm}) \approx (P_{eq} - 1)V_r(T, 0)$$

At equilibrium, $\mu_m(T, P_{eq}) = \mu_r(T, P_{eq})$ and so

$$-[(\mu_m(T, P = 1 \text{ atm}) - \mu_r(T, P = 1 \text{ atm})] = (P_{eq} - 1)(V_m - V_r)$$

or

$$-\Delta\mu(T, P = 1 \text{ atm}) = (P_{eq} - 1)\Delta V$$

which is the same as outlined in *point 1*, above.

(3) $(\partial \ln K_a/\partial P)_T = -\Delta V^0/RT \rightarrow \ln K_a(T, P_{eq}) - \ln K_A(T, P = 1 \text{ atm})$
$\approx -(\Delta V^0/RT)(P_{eq} - 1)$, $K_a(T, P_{eq}) = 1$, and thus $-\ln K_a(T, P = 1 \text{ atm}) = -\Delta V^0/RT(P_{eq} - 1)$. Because $\ln K_a(T, P = 1\text{atm}) = -\mu^0/RT$, we obtain

$$\Delta\mu^0(T, P = 1 \text{ atm}) = -\Delta V^0(P_{eq} - 1)$$

which is the same as outlined in *point 1* above.

2. $\overline{H}(T, P) = \int^T \overline{C}_{P\rightarrow 0}(T)dT' + \overline{H}_0 + \int_0^P [\overline{V} - T(\partial V/\partial T)_{P'}]dP'$

$$\overline{V} = RT/P + B_2(T); T(\partial V/\partial T)_{P'} = RT/P' + T \, dB_2/dT$$

$$\overline{H}(T, P) = \int^T C_{P\rightarrow}(T)dT' + H_0 + [B_2(T) - T \, dB_2/dT]P$$

$B_2(T) = b - a/RT; T\, dB_2/dT = aT/RT^2$

$$\rightarrow \overline{H}(T,P) = \int^T C_{P\rightarrow 0}\, dT' + \overline{H}_0 + [b - 2a/RT]P$$

Then for *problems 2a* and *2b*, respectively,

(a) $\overline{C}_P(T,P) = (\partial H/\partial T)_P = \overline{C}_{P\rightarrow 0}(T)$
$\qquad\qquad + [(dB_2/dT - T d^2B_2/dT^2 - dB_2/dT)]P$
$\qquad = \overline{C}_{P\rightarrow 0}(T) - T(d^2B_2/dT^2)P$

(b) $\overline{C}_P(T,P) = \overline{C}_{P\rightarrow 0}(T) + (2a/RT^2)P$

3. When $P^l = 1$ atm and $P^v = 1$ atm, water boils at 100°C or 373 K. When $P^l = 10^3$ atm and $P^v = 1$ atm, the boiling temperature, T_b, can be obtained from the generalized Clapeyron Equation:

$(\partial P^l/\partial T)_{P^v} = -\Delta\overline{H}_{vap}/T\overline{V}^l$

$\qquad = -540\,\text{cal}/(T \times 1\,\text{cm}^3) \times 82.06 \times 10^4\,\text{atm cm}^3/1.987\,\text{cal}$

$\qquad \approx -2.23 \times 10^4\,\text{atm}/T$

$\int_1^{1,000} dP' \approx -2.23\,10^4\,\text{atm} \int_{373}^{T_b} dT/T = -2.23 \times 10^4\,\text{atm}\ln(T_b/373\,\text{K})$

$(1,000 - 1) \approx -2.23 \times 10^4\,\text{atm}\ln(T_b/373)$

$\qquad T_b \approx 357\,\text{K}$

4. For an ideally dilute solution, $\mu_i = \mu_i^0 + RT\ln x_i$

$$\begin{array}{lll} \text{Con I}: & \mu_i^0 = \mu_i^\bullet & \text{(all } i) \\ \text{Con II}: & \mu_1^0 = \mu_1^\bullet & \text{(solvent ``1'')} \\ & \mu_i^0 \neq \mu_i^\bullet & \text{(solute ``}i\text{''}; i \neq 1) \end{array}$$

By definition, $\Delta G_{mix} = \sum_i n_i(\mu_i^0 - \mu_i^\bullet)$.

- Con I: $\Delta G_{mix} = RT\, \Sigma n_i \ln x_i$ $\qquad$ (all i)

$\qquad\qquad \Delta S_{mix} = -(\partial\Delta G_{mix}/\partial T)_P = -R\ln n_i \ln x_i$

$\qquad\qquad \Delta H_{mix} = [\partial(\Delta G_{mix}/T)/\partial(1/T)]_P = 0$

$\qquad\qquad \Delta V_{mix} = (\partial\Delta G_{mix}/\partial P)_T = 0$

- Con II:

$$\Delta G_{mix} = \Sigma_i n_i(\mu_i - \mu_i^0) + n_1(\mu_1^0 - \mu_1^\bullet) + \Sigma_{i\neq 1} n_i(\mu_i^0 - \mu_i^\bullet)$$
$$= RT\Sigma_i n_i \ln x_i + 0 + \Sigma_{i\neq 1} n_i(\mu_i^0 - \mu_i^\bullet)$$
$$\Delta S_{mix} = -R\Sigma_i \ln x_i - \Sigma_{i\neq 1} n_i(S_i^0 - S_i^\bullet)$$
$$\Delta H_{mix} = 0 + \Sigma_{i\neq 1} n_i(H_i^0 - H_i^\bullet)$$
$$\Delta V_{mix} = 0 + \Sigma_{i\neq 1} n_i(V_i^0 - V_i^\bullet)$$

SOLUTION TO SET VIII

1. $$(\partial T/\partial\mathcal{A})_S = -(\partial S/\partial\mathcal{A})_T/(\partial S/\partial T)_A = (T/C_A)(\partial\sigma/\partial T)_A$$
$$= \sigma_0 n(1 - T/T_c)^{n-1}(-1/T_c) < 0$$

2a. For $dE = TdS - PdV + Ydy + \mu dn; E = TS - PV + Yy + \mu n$:
$$\psi = E - V(\partial E/\partial V)_{S,y,n} = E + PV$$
$$d\psi = dE + PdV + VdP = TdS - PdV + Ydy + \mu dn + PdV + VdP$$
$$= TdS + VdP + Ydy + \mu dn$$

2b. $$\phi = E - S(\partial E/\partial S)_{V,y;\underline{n}_i} - V(\partial E/\partial V)_{S,y,\underline{n}_i} - y(\partial E/\partial y)_{S,V;\underline{n}_i}$$
$$= E - TS + PV - Yy$$
$$d\phi = dE - TdS - SdT + PdV + VdP - Ydy - ydY$$
$$= -SdT + VdP - ydY + \mu dn$$

3a. Since T is held constant, it is natural to use the Helmholtz Free Energy, A, i.e., $dA = -SdT + fdL$ (assuming PV is constant).

- Conditions for equilibrium: $\delta^{(1)}A_{T,L} = 0$
- Conditions for stability: $\delta^{(2)}A_{T,L} > 0$

Divide strip into two parts, $L^{(1)}$ and $L^{(2)}$. Because L is constant, $\delta L^{(1)} = -\delta L^{(2)}$. (Note that T is fixed and uniform throughout.)

$$\delta^{(1)}A_{T,L} = -f^{(1)}dL^{(1)} + f^{(2)}dL^{(2)} = (f^{(1)} - f^{(2)})\delta L^{(1)} = 0 \Rightarrow f^{(1)} = f^{(2)} = f$$
$$\delta^{(2)}A_{T,L} = \frac{1}{2}\{[\partial^2 A/\partial(L^{(1)})^2]_{T,L^{(2)}}(\delta L^{(1)})^2 + [\partial^2 A/\partial(L^{(2)})^2]_{T,L^{(1)}}(\delta L^{(2)})^2\} > 0$$

Since $(\delta L^{(1)})^2 = (-\partial L^{(2)})^2$ we have $(\partial^2 A/\partial L^2)_T > 0$ or $(\partial f/\partial L)_T > 0$.

3b. If $f = T\phi(L)$, then to satisfy *solution 3a*, we must have $(\partial f/\partial L)_T = Td\phi/dL > 0$, i.e., $\phi(L)$ must be a monotonically increasing function of L.

3c. $\alpha = 1/L(\partial L/\partial T)_f = -1/L(\partial f/\partial T)_L/(\partial f/\partial L)_T$

$\quad = -(1/L)\phi(L)/T(d\phi/dL) < 0$

3d. $(\partial E/\partial L)_T = f - T(\partial f/\partial T)_L = T\phi(L) - T\phi(L) = 0$

3e. $dE = dq + dw = dq + fdL; \Delta E = q + \int fdL; q = \Delta E - \int fdL; \Delta E = 0$ at constant T; $q = -\int(fdL) = -T\int_1^2 \phi(L)dL < 0$ since ϕ is an increasing function of L.

4. $(\partial^2 A/\partial V^2)_{T,\underline{n}_i} = -(\partial P/\partial V)_{T,\underline{n}_i} \geq 0$

$\quad (\partial^2 G/\partial P^2)_{T,\underline{n}_i} = -(\partial V/\partial P)_{T,\underline{n}_i} = 1/(\partial P/\partial V)_{T,n_i} \leq 0$

$\quad (\partial^2 E/\partial S^2)_{V,\underline{n}_i} = (\partial T/\partial S)_{V,\underline{n}_i} = T/C_V \geq 0$

$\quad (\partial^2 A/\partial T^2)_{V,\underline{n}_i} = -(\partial S/\partial T)_{V,\underline{n}_i} = C_V/T \leq 0$

SOLUTION TO SET IX

1. In one molecule there are two atoms of class ^{35}Cl and class ^{37}Cl. The abundance of ^{35}Cl is 0.75% and of ^{37}Cl is 0.25%. This gives rise to the following probabilities:

$$^{35}Cl_2 = [2!/(2!0!)](0.75)^2 = 0.5635$$

$$^{35}Cl - {}^{37}Cl = [2!/(1!1!)](0.75)(0.25) = 0.375$$

$$^{37}Cl_2 = [2!/(0!2!)](0.25)^2 = 0.0625$$

2.

$$\Omega_D^{CMB} = \Pi_i g_i^{N_i}/N_i!; \ln \Omega_D^{CMB}$$

$$= \Sigma_i(N_i \ln g_i - N_i \ln N_i + N_i)$$

$$\ln \Omega_D^{CMB} = [N_1 \ln g_1 + \ldots N_i \ln g_i \ldots - (N_1 \ln N_1 + \ldots N_i \ln N_i + \ldots)$$
$$+ (N_1 + \ldots N_i + \ldots)]$$

Differentiating with respect to N_i, which is regarded here as a particular variable, holding all other $N_{j \neq i}$ constant, gives

$$\partial \ln \Omega_D^{CMB}/\partial N_i = \ln g_i - N_i/N_i - \ln N_i + 1 = \alpha + \beta\varepsilon_i$$

Thus, $\ln g_i/N_i = \alpha + \beta\varepsilon_i$ or $N_i = g_i e^{-\alpha}e^{-\beta\varepsilon_i}$ and therefore $N = e^{-\alpha}\Sigma_i g_i e^{-\beta\varepsilon_i}; e^{-\alpha} = g_i e^{-\beta\varepsilon_i}/\Sigma_i g_i e^{-\beta\varepsilon_i}$ and $N_i = Ng_i e^{-\beta\varepsilon_i}/\Sigma_i g_i e^{-\beta\varepsilon_i}$.

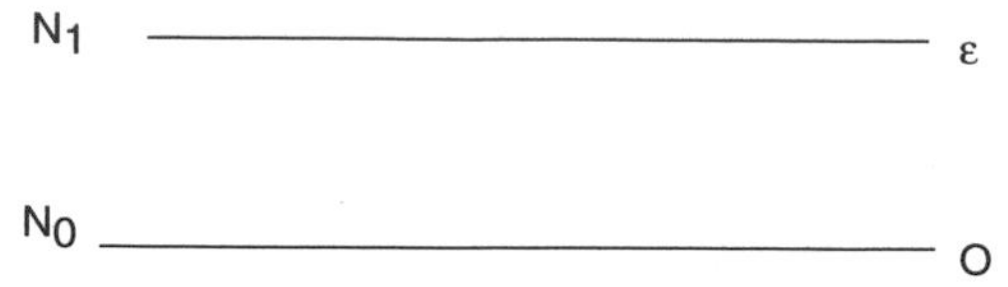

$$N_1 \rule{6cm}{0.4pt} \quad \varepsilon$$

$$N_0 \rule{7cm}{0.4pt} \quad 0$$

Figure S9.1   Two energy levels

3. (See Fig. S9.1) $N_i/N = g_i e^{-\beta \varepsilon_i}/\Sigma_i g_i e^{-\beta \varepsilon_i}; N_1/1{,}000$

$$= 3e^{-\beta}/(3e^{-\beta} + 2e^{-2\beta})$$

$$N_2/1{,}000 = 2e^{-2\beta}/(3e^{-\beta} + 2e^{-2\beta})$$

$$E = N_1 \varepsilon_1 + N_2 \varepsilon_2$$

$$= [1{,}000/(3e^{-\beta} + 2e^{-2\beta})]$$

$$\times [3e^{-\beta} \times 1 + 2e^{-2\beta} \times 2]$$

Solve for β: $= 1{,}200$

$$10(3e^{-\beta} + 4e^{-2\beta}) = 12(3e^{-\beta} + 2e^{-2\beta})$$

$$30e^{-\beta} + 40e^{-2\beta} = 36e^{-\beta} + 24e^{-2\beta} \Rightarrow 30 + 40e^{-\beta}$$

$$= 36 + 24e^{-\beta} \Rightarrow e^{-\beta} = 3/8.$$

Therefore,

$$N_1 = [1{,}000(3)(3/8)]/[3(3/8) + 2(3/8)^2] = 1{,}000/(1 + 0.25) = 800$$
$$N_2 = [1{,}000(2)(3/8)^2]/[3(3/8) + 2(3/8)^2] = 1{,}000/5 = 200.$$

4. $N_1/N = e^{-\beta \varepsilon}/[e^{-\beta \varepsilon x 0} + e^{-\beta \varepsilon}] = e^{-\beta \varepsilon}/[1 + e^{-\beta \varepsilon}] = 1/[e^{\beta \varepsilon} + 1]$

Thus,
$N_1 e^{\beta \varepsilon} + N_1 = N; e^{\beta \varepsilon} = (N - N_1)/N_1; \beta = (1/\varepsilon) \ln[(N - N_1)/N_1]$
since $\beta = 1/kT$ and $T = \varepsilon/\{k \ln[(N - N_1)/N_1]\}$. Observing that
$E = \Sigma_i N_i \varepsilon_i = N_0 \times 0 + N_1 \times \varepsilon = N_1 \varepsilon$, it follows that

when $E_> \to 0$	$N_1 \to 0$	$\ln[(N - N_1)/N_1] \to \infty_>$	and $T \to 0$
when $E_< \to \frac{1}{2}N_1 \varepsilon$	$N_1 \to \frac{1}{2}N_<$	$\ln[(N - N_1)/N_1] \to 0_>$	and $T \to \infty$
when $E_> \to \frac{1}{2}N_1 \varepsilon$	$N_1 \to \frac{1}{2}N_>$	$\ln[(N - N_1)/N_1] \to 0_<$	and $T \to -\infty$
when $E_< \to N\varepsilon$	$N_1 \to N_<$	$\ln[(N - N_1)/N_1] \to \infty_<$	and $T \to 0$

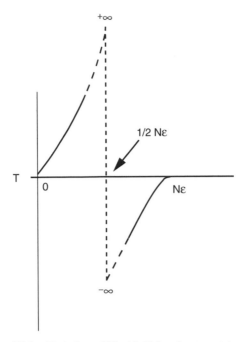

Figure S9.2   Variation of T with E for the two-state system

The symbols $>$ and $<$ mean that the value is slightly larger or slightly smaller than the quantity to which they refer. A plot of T vs. E looks like the graph in Fig. S9.2. As E increases from 0 to Nε, the temperature (the absolute temperature!) increases from 0 to infinity, changes abruptly from $+\infty$ to $-\infty$ at $1/2$Nε, and then continues with negative values until T = 0 is reached at E = Nε.

Note: This peculiar behavior of the temperature comes about because of the assumption of the presence of a finite number of states. We only used two states in this example; if we had used a larger but finite number of states, we would have arrived at a similar conclusion. In a realistic system, there are an infinite number of states and as the energy rises these states become activated and no negative absolute temperatures are predicted. Nonetheless, if the system consists of several modes of motion and one of these modes has a finite number of states (as, for example, magnetic states) that remain essentially independent for a long time from the rest of the system, then such a system can give rise to absolute negative temperatures.

SOLUTION TO SET X

1. $\nu_o = 6.98 \times 10^{13} \text{ s}^{-1}$;

$N_1/N_0 = exp[-(1/2+1)h\nu_o/kT]/exp[(-1/2h\nu_o+0)h\nu_o/kT)] = exp(-h\nu_o/kT)$

$N_1/N_2 = exp[-(6.626 \times 10^{-27} \times 6.98 \times 10^{13})/1.38 \times 10^{-16} \times T] = e^{-3351.44/T}$

$$\frac{N_1/N_0}{}$$

(a) $T = 25°C = 298 \text{ K}$ 1.31×10^{-5}
(b) $T = 800°C = 1,073 \text{ K}$ 4.4×10^{-2}
(c) $T = 3,000°C = 3,273 \text{ K}$ 0.36

2. $S = k \ln \Omega_D^{CMB} = k \ln \Pi_i(g_i/N_i!) = k\Sigma_i(N_i \ln g_i - N_i \ln N_i + N_i)$
    $= k\Sigma_i(\ln g_i/N_i + 1)$

Because $N_i/N = g_i e^{-\beta\varepsilon_i}/q$ or $g_i/N_i = q/Ne^{\beta\varepsilon_i}$, we obtain

$S = k\Sigma_i N_i[\ln q/N + \beta\varepsilon_i + 1) = kN \ln q/N + \beta k\Sigma_i N_i \varepsilon_i + kN$. Substitute $\beta = 1/kT$, $\Sigma_i N_i \varepsilon_i = E = kNT^2(\partial \ln q/\partial T)_{V,N}$ and $S = kN \ln(q/N) + kN$ $(\partial \ln q/\partial T)_{V,N} + kN$.

- Effect of the "zero" of energy on S:

$$\varepsilon_i' = \varepsilon_i + \varepsilon_i$$

$$q' = \Sigma_i g_i e^{-\beta\varepsilon_i'} = (\Sigma_i g_i e^{-\beta\varepsilon_i})e^{-\beta\varepsilon_0} = qe^{-\beta\varepsilon_0}$$

$$S = kN \ln q'/N + \beta k\Sigma_i N_i \varepsilon_i' + kN$$

$$= kN \ln q/N - kN\beta\varepsilon_0 + \beta k\Sigma_i N_i \varepsilon_i + \beta kN\varepsilon_0 + kN$$

The zero of energy terms cancel, and the remaing expression is the same as above without the zero of energy. Thus, S is unaffected by the zero of energy.

3. He is monatomic. There are no vibrational and rotatational contributions. There are also no electronic contributions, since we are only considering the ground state, and $g_{el,0} = 1$. Thus, the total partition function is just the translational partition function.

$$E - E_0 = kNT^2(\partial \ln q/\partial T)_V; q_{tr}$$

$$= (2\pi mkT/h^2)^{3/2} V; \ln q_{tr} = 3/2 \ln T + \text{const}$$

$$N = N_0 = 6.02 \times 10^{23}; T = 273 \text{ K};$$

$$E - E_0 = 3/2 \times 1.38 \times 10^{-16} \, \text{erg} \cdot \text{K}^{-1} \cdot \text{molecules}^{-1}$$

$$\times 6.02 \times 10^{23} \, \text{molecules/mol} \times 273 \, \text{K}$$

$$= 3.402 \times 10^{10} \, \text{erg/mol} = 3.402 \times 10^3 \, \text{J/mol}$$

$$H - E_0 = kNT^2 (\partial \ln q / \partial T)_{V,N} + (NkT/V) = 3/2 \, NkT + NkT$$

$$= 5/2 NkT = 5.669 \, 10^{10} \, \text{erg/mol} = 5.669 \times 10^3 \, \text{J/mol}$$

$$A - E_0 = -kNT \ln q / N - kNT = -kNT(\ln q / N + 1)$$

$$= -1.38 \times 10^{-16} \times 6.02 \times 10^{23} \times 273 \{ \ln[2\pi \times (4/6.02 \times 10^{23})$$

$$\times (1.38 \times 10^{-16} \times 273/(6.626 \times 10^{-27})^2]^{3/2}$$

$$\times 22.4 \times 10^3 / 6.02 \times 10^{23}] \, \text{erg/mol}$$

$$= -3.047 \times 10^{11} \, \text{erg/mol} = -3.047 \times 10^4 \, \text{J/mol}$$

$$\mu - \varepsilon_0 = -kT \ln q / N = (A - E_0)/(6.02 \times 10^{23}) + kT$$

$$= -3.047 \times 10^{11}/6.02 \times 10^{23} = 1.38 \times 10^{-16} \times 273 \, \text{erg/molecules}$$

$$= -4.686 \, 10^{-13} \, \text{erg/molecules} = -4.686 \times 10^{-20} \, \text{J/molecules}$$

$$C_V = (\partial E / \partial T)_V = 3/2 \times 1.38 \, 10^{-16} \times 6.02 \times 10^{23}$$

$$= 1.246 \times 10^8 \, \text{erg} \cdot \text{K}^{-1} \, \text{mol}^{-1} = 12.46 \, \text{J} \cdot \text{K}^{-1} \cdot \text{mol}^{-1}$$

$$(= 2.98 \, \text{cal} \cdot \text{K}^{-1} \cdot \text{mol}^{-1})$$

4.
$$K_c = \Pi_i (N_i/V)^{v_i} = \Pi_I (q_i/V)^{v_i} e^{-\Delta \varepsilon_0 kT}$$

$$Cl_2 \Leftrightarrow 2 \, Cl$$

$$K_c = [(q_{cl}/V)^2]/(q_{Cl_2}/V)] e^{-(2\varepsilon_{0,Cl} - \varepsilon_{0,Cl_2})/kT}; \varepsilon_{0,Cl} = 0; \varepsilon_{0,Cl_2} = -D_0$$

$$D_0 = 238.9 \times 10^3 \, \text{J/mol} = 238.9 \times 10^{10} \, \text{erg/mol}$$

$$= (238.9 \, 10^{10}/6.02 \times 10^{23}) \, \text{erg/molecule}$$

$$= 3.968 \times 10^{-12} \, \text{erg/molecule}$$

$$(q_{Cl}/V)^2 = \{ [(2\pi \times 35/6.02 \times 10^{23}) \times 1.38 \times 10^{-16}$$

$$\times 2,000/(6.63 \times 10^{-27})^2]^{3/2} \times 4 \}^2$$

$$= 1.388 \times 10^{28} \, \text{molecules}^2 \cdot \text{cm}^{-6}$$

Note: q is dimentionless, but q/V is not. In the cgs system, which we are using here, it has the dimensions of (volume per molecule)$^{-1}$, i.e., q/V has the dimensions of molecule/cm^3/cm^{-3}.

$$(q_{Cl_2}/V) = [2\pi \times (70/6.02 \times 10^{23}) \times 1.38 \times 10^{-16} \times 2,000/(6.63\,10^{-27})^2]^{3/2}$$
$$\times\; [2,000/(2 \times 0.351)] \times [1/(1 - e^{-813/2000})]$$
$$= 9.818 \times 10^{27} \times 2.849 \times 10^3 \times 2.99$$
$$= 8.374 \times 10^{31}\,\text{molecules/cm}^3$$

$$e^{-\Delta\varepsilon_0/kT} = exp(\varepsilon_{0,Cl_2}/kT) = e^{-D_0/kT}$$
$$= \exp(-3.968 \times 10^{-12}/1.38 \times 10^{-16} \times 2000)$$
$$= 5.696 \times 10^{-7}$$

$$K_c = [(1.388 \times 10^{28})^2 \times 5.696 \times 10^{-7}]/$$
$$\times\; [9.818 \times 10^{27} \times 2.849 \times 10^3 \times 2.99]$$
$$= 1.312 \times 10^{18}\,\text{molecules/cm}^3 = 2.18 \times 10^{-6}\,\text{mol/cm}^3$$

INDEX

Thermodynamics and Introductory Statistical Mechanics, by Bruno Linder
ISBN 0-471-47459-2 © 2004 John Wiley & Sons, Inc.